Geography of the Physical Environment

The *Geography of the Physical Environment* book series provides a platform for scientific contributions in the field of Physical Geography and its subdisciplines. It publishes a broad portfolio of scientific books covering case studies, theoretical and applied approaches as well as novel developments and techniques in the field. The scope is not limited to a certain spatial scale and can cover local and regional to continental and global facets. Books with strong regional focus should be well illustrated including significant maps and meaningful figures to be potentially used as field guides and standard references for the respective area.

The series appeals to scientists and students in the field of geography as well as regional scientists, landscape planners, policy makers, and everyone interested in wide-ranging aspects of modern Physical Geography. Peer-reviewed research monographs, edited volumes, advance and undergraduate level textbooks, and conference proceedings covering the major topics in Physical Geography are included in the series. Submissions to the Book Series are also invited on the theme 'The Physical Geography of…', with a relevant subtitle of the author's/editor's choice. Please contact the Publisher for further information and to receive a Book Proposal Form.

More information about this series at https://link.springer.com/bookseries/15117

Shruti Kanga · Gowhar Meraj · Majid Farooq ·
Suraj Kumar Singh · Mahendra Singh Nathawat
Editors

Disaster Management in the Complex Himalayan Terrains

Natural Hazard Management, Methodologies and Policy Implications

Springer

Editors
Shruti Kanga
Centre for Climate Change and Water Research
Suresh Gyan Vihar University
Jaipur, Rajasthan, India

Majid Farooq
Department of Ecology, Environment and Remote Sensing
Government of Jammu and Kashmir
Srinagar, Jammu and Kashmir, India

Mahendra Singh Nathawat
Department of Geography
Indira Gandhi National Open University
New Delhi, Delhi, India

Gowhar Meraj
Department of Ecology, Environment and Remote Sensing
Government of Jammu and Kashmir
Srinagar, Jammu and Kashmir, India

Suraj Kumar Singh
Centre for Sustainable Development
Suresh Gyan Vihar University
Jaipur, Rajasthan, India

ISSN 2366-8865 ISSN 2366-8873 (electronic)
Geography of the Physical Environment
ISBN 978-3-030-89307-1 ISBN 978-3-030-89308-8 (eBook)
https://doi.org/10.1007/978-3-030-89308-8

Cover image by Sonja Weber, München

This Springer imprint is published by the registered company Springer Nature Switzerland AG
The registered company address is: Gewerbestrasse 11, 6330 Cham, Switzerland

Preface and Acknowledgements

The present book is an outcome of the collaboration between Suresh Gyan Vihar University, Jaipur, Department of Ecology, Environment and Remote Sensing, Government of Jammu and Kashmir, and the Department of Geography, Indira Gandhi National Open University, New Delhi, India. The book was conceptualized during the International Conference on Climate Change and Water Research 2020, Suresh Gyan Vihar University, when the editors discussed anthropogenic factors augmenting the risk in the Himalayan region toward the natural hazards. From that day onward, we put a lot of effort and motivation into making this book become a reality. This book discusses different types of hazards prevalent in the Himalayan region, their evolution, science, and mitigation. The book has tried to collate all the significant natural hazards at one place discussing their genesis and how mitigation can be made effective if policies are designed considering the development and sustainability together. Under these paradigms, the hazards can be rightly said that they are manageable in the context of reducing the deleterious impacts on the environment and the loss of human life and property. This book aims to provide a comprehensive discussion on the evolution, mitigation, and vulnerability in the face of natural hazards, emphasizing the role of satellite remote sensing technology. The book comprehensively discusses the hazards in the Himalayan terrain's perspective through a systematic review of the current literature. Natural hazards have always been regarded as the destroyer of economies. More specifically, the floods in the rugged Himalayan terrains have washed away numerous hydel power projects. One of the case studies of our book has focused on this issue. The role of Glacial Lake Outburst Floods (GLOFs) has also been described in detail, with a particular focus on how they can be modeled and mapped. In September 2014, Kashmir Valley, India's northernmost region, witnessed one of the most devastating floods in the century. A particular focus on its genesis and evolution has also been discussed.

Landslides are a widespread phenomenon in the Himalayas that has got aggravated due to highway constructions. Special chapters were dedicated to discussing them in detail. Moreover, earthquakes are part and parcel of people living in the Himalayas, and what if humans can forecast them with some certainty, that would reduce the loss of life in these regions. The forecasting of earthquakes has also been discussed in detail with some innovative techniques to make it happen. Further, we have also discussed the Government of India's approach to tackling forest fire and Jammu and Kashmir's case for assessing the region's fire vulnerability. A special chapter has been dedicated to remote sensing and GIS for assessing the natural hazards in the Himalayas. Also, how policies can be used to mitigate the natural risks in the Himalayas have been discussed in one of the chapters.

Disaster Management in the Complex Himalayan Terrains: Natural Hazard Management, Methodologies and Policy Implications is a comprehensive edited book focusing on managing natural hazards using innovative techniques of spatial information sciences and satellite remote sensing. Chapters have been written by eminent researchers and experts in the field of hazard management, remote sensing, and GIS with a focus to replenish the gap in the available literature on the subject by bringing the concepts, theories, and practical experiences of the specialists and professionals in this field together in one volume. The book shall help students, researchers, and policymakers develop a complete understanding of the management and policy implications of natural hazards in the complex Himalayan region.

We hope that through this book project, the team of authors will become part of an academic community working specifically on the mitigation and management of the natural hazards on the Himalayas using spatial sciences and state of the art of modeling tools. We want to acknowledge the help of all the reviewers who tirelessly read chapters and sent suggestions to authors that greatly enhanced their quality and prospective reach. Special thanks are to Dr. Neelu Gera, Dr. Muhammad Muslim, and Dr. Muzamil Amin, for their valuable suggestions during the book's proofreading. Finally, we would like to thank our families who supported us in thick and thin at each stage of our lives; nothing would have been possible without their help and support. We hope you shall enjoy reading this book on the various perspectives of the natural disaster–human interference relationship. Unfortunately, there are almost certainly more approaches than we could include in the following chapters, and a number of these may leave you with more questions than answers. However, we believe that the chapters featured in this book will advance our discussion on the genesis and role of humans in the occurrence of natural hazards in the Himalayas.

Editors

Jaipur, India — Shruti Kanga
Kashmir, India — Gowhar Meraj
Kashmir, India — Majid Farooq
Jaipur, India — Suraj Kumar Singh
New Delhi, India — Mahendra Singh Nathawat

Contents

Introduction to the Disasters and Hazards in the Himalayan Terrains

Disasters in the Complex Himalayan Terrains 3
Shruti Kanga, Gowhar Meraj, Majid Farooq, Suraj Kumar Singh, and Mahendra Singh Nathawat

Hazards in the Perspective of Himalayan Terrain: A Review 11
Fayma Mushtaq, Afaan Gulzar Mantoo, Anamika Shalini Tirkey, and Sofi Zubair Ahmad

Landslides in the Himalayas—Causes and Consequences

Himalayan Landslides–Causes and Evolution 33
Sandeep Singh, Anand Joshi, Anamika Sahu, R. Arun Prasath, Saurabh Sharma, and Chandra Shekhar Dwivedi

Landslides in the Himalayas: Causes, Evolution, and Mitigation—A Case Study of National Highway 44, India 43
Mohsin Fayaz, Sheik Abdul Khader, and Mohammd Rafiq

Geo-Environmental Impact of Road Widening Project Along National Highway-44, Jammu and Kashmir, India 59
Vinay Kumar Pandey

Floods in the Himalayas—Causes and Consequences

September 2014 Floods in Kashmir Himalaya—Impacts and Mitigation Strategy 81
Mohammd Rafiq, Amit Parashuram Kesarkar, Usamah Derwaish, and Aaqib Manzoor Bhat

A Review on the Estimation of Glacial Lake Outburst Floods (GLOFs) in the Himalayan Region Using Remote Sensing and Geographic Information System ... 93
Mohd Talha Anees, Naseem Akhtar, Ahmad Farid Bin Abu Bakar, and Muhammad Izzuddin Syakir Ishak

Impact of Floods on the Green Energy Sector in Himalayas—A Case Study of Gagas Watershed, Uttarakhand, India ... 109
Sapna Bisht, Smita Chaudhry, Subrat Sharma, and Surajit Dutta

Earthquakes in the Himalayas—Assessment and Forecasting

Synthetic Aperture Radar Interferometry to Measure Earthquake-Related Deformation: A Case Study from Nepal ... 133
Himanshu Verma, Yogendra Sharma, and Sumanta Pasari

Earthquake Forecasting in the Himalayas Artificial Neural Networks ... 141
Arnav Ahuja and Sumanta Pasari

Hazard Mitigation Strategies in the Himalayas

Forest Fire Alert System of India with a Special Reference to Fire Vulnerability Assessment of the UT of Jammu and Kashmir ... 155
Majid Farooq, Soheib Gazali, Mudasir Dada, Neelu Gera, and Gowhar Meraj

Hazard Mitigation and Climate Change in the Himalayas–Policy and Decision Making ... 169
Mohammd Rafiq, Gowhar Meraj, Amit Parashuram Kesarkar, Majid Farooq, Suraj Kumar Singh, and Shruti Kanga

About the Editors

Dr. Shruti Kanga, M.Sc., Ph.D. is currently working as Associate Professor and Coordinator, Centre for Climate Change and Water Resources, Suresh Gyan Vihar University, Jaipur. She has previously worked as Assistant Professor in the Centre for Land Resource Management, Central University of Jharkhand (CUJ), Ranchi. She had also served as Research Associate in Ministry of Environment, Forest and Climate Change (MoEF), GoI-funded project on Forest Fires, and has teaching experience in different universities, i.e., Central University of HNB Garhwal and the University of Jammu. She did her Ph.D. degree in technology (geoinformatics) from the Department of Remote Sensing, Birla Institute of Technology, Ranchi, in 2013. She has worked in the area of forest fires risk modeling and management, tourism, resource management, etc. She has attended and organized around 35 national and international conferences. She was also Course Coordinator of different modules of EDUSAT-based distance learning program organized by Indian Institute of Remote Sensing (IIRS), Indian Space Research Organisation (ISRO), Government of India. I have 12 years of teaching and research experience.
Google Scholar profile: https://scholar.google.co.in/citations?user=RSkmve8AAAAJ&hl=en&oi=ao

Dr. Gowhar Meraj, M.Sc., M.Phil., Ph.D. is currently working as Young Scientist Fellow (YSF), under the Department of Science and Technology, Government of India's Scheme for Young Scientists and Technologists (SYST-SEED) at the Department of Ecology, Environment and Remote Sensing, Government of Jammu and Kashmir. Previously, he has worked as Programme Officer on ENVIS Scheme of Ministry of Environment, Forests, and Climate Change (MoEF & CC) for five years. He has also worked as a consultant with the World Bank Group, New Delhi, for its South Asia Water Initiative (SAWI) Program.

He has done his M.Sc., M.Phil., and Ph.D. in Environmental sciences with research focus on the application of spatial information sciences in policy and decision making in the Himalayan region. He has more than 10 years of experience in research, teaching, and consultancy in the fields of satellite remote sensing, geographic information system (GIS), hazard vulnerability assessments, hydrological modeling,

flood assessment, watershed management, natural resource management, ecosystem services, and modeling. He has more than 50 publications in various reputed peer-reviewed journals. He is on editorial and reviewer boards of various WoS/SCI journals such as *Information Fusion, Natural Hazards, Arabian Journal of Geosciences, Geocarto International, and various others.*
Google Scholar profile: https://scholar.google.co.in/citations?user=tLe6BeEAAAAJ&hl=en

Dr. Majid Farooq, M.Sc., M.Tech., Ph.D. is currently working as Scientist-D in the Department of Ecology, Environment and Remote Sensing, Government of Jammu and Kashmir, India. He has more than 15 years of experience in research, teaching, and consultancy related to remote sensing and GIS such as climate change vulnerability assessments, flood modeling, ecosystem assessment, watershed management, natural resource management, ecosystem services, and modeling.

He has done his M.Tech. and Ph.D. in remote sensing and GIS and M.Sc. in Environment and Ecology. He has more than 40 publications in various reputed peer-reviewed journals. Besides, he is also serving as reviewer in various SCI journals such as *Environmental Monitoring and Assessment* (Springer), *International Journal of Remote Sensing* (Taylor & Francis) and various others.
Google Scholar profile: https://scholar.google.co.in/citations?user=6X2n7iMAAAAJ&hl=en

Dr. Suraj Kumar Singh, M.Sc., Ph.D. is currently working as Associate Professor and Coordinator, Centre for Sustainable Development, Suresh Gyan Vihar University, Jaipur. Previously, he had worked as Assistant Professor in the Centre for Land Resource Management, Central University of Jharkhand (CUJ), Ranchi. He has also served as Research Associate in National Remote Sensing Centre (NRSC), Indian Space Research Organisation (ISRO), Government of India. Sponsored projects, i.e., National Wasteland Mapping and National Urban Information System for Northern parts of India. He did his Ph.D. degree in technology (geoinformatics) from the Department of Remote Sensing, Birla Institute of Technology, Ranchi, in 2012. He has worked in the area of waterlogging and flood hazards, geospatial applications in water resources, disaster management, hydrogeomorphology, urban planning, and wasteland mapping. He was also Course Coordinator of different modules of EDUSAT-based distance learning program organized by Indian Institute of Remote Sensing (IIRS), Indian Space Research Organisation (ISRO), Government of India. He has 12 years of teaching and research experience. He had the responsibility of NSS Program Officer at the Central University of Jharkhand and Coordinator of Centre for Sustainable Development, Suresh Gyan Vihar University, Jaipur.
Google Scholar profile: https://scholar.google.co.in/citations?user=wXlG45AAAAAJ&hl=en

Prof. Mahendra Singh Nathawat is a well-known Geographer. After his M.A. in Geography from Rajasthan University (1980), he worked in South Asia Study Centre, University of Rajasthan as a Research Fellow (1982-1984), and then he joined as a Scientist/Lecturer (1985-93) and Senior Scientific Officer and in-charge (1993-98) in Dept. of Remote Sensing, B. M. Birla Science & Technology Centre, Jaipur. He also joined State NRDMS Centre, Dept. of Science & Technology, Govt. of Haryana as a Project Director (1998-2001). Prof. Nathawat headed the Department of Remote Sensing & Geoinformatics, Birla Institute of Technology (BIT), Mesra, Ranchi (India), from 2001 to 2011. Currently, Prof. Nathawat has been working as a Professor in Geography Discipline School of Sciences, Indira Gandhi National Open University (IGNOU), New Delhi, India, since 2011. He did his PG Diploma (1986-87) in Remote Sensing Techniques in Geology & Geomorphology, Indian Institute of Remote Sensing, Dept. of Space, Dehradun. He has completed his Ph.D. (1992-96) at the University of Rajasthan in Geomorphological Aspects of the Processes of Desertification in Rajasthan: Monitoring & Dynamics with Remote Sensing. His area of specialization and research interest are Desertification, GIS, Natural Resources Management, Regional & Urban Planning, Environmental Geomorphology, Climate Change, Sustainable livelihood, and Disaster Management. He has 30 years of research experience and 16 years of teaching experience in Remote Sensing and GIS. He has published more than 60 research papers in various international and national journals and written eight books. Prof Nathawat also made significant contributions in the number of Atlases published by NRSC (ISRO), HSCST, SAC, etc. He has been awarded Baden-Wuerttemberg Fellowship-2009-10 (Under Indo-German Cultural Exchange Programme). Under his supervision, more than 15 PhD. students and 42 M.Tech./M.Sc. students have completed their degree.

Introduction to the Disasters and Hazards in the Himalayan Terrains

Disasters in the Complex Himalayan Terrains

Shruti Kanga, Gowhar Meraj, Majid Farooq, Suraj Kumar Singh, and Mahendra Singh Nathawat

Abstract Understanding the science behind the causal factors of natural hazards in the Himalayan region is emerging. Hazards, disasters, associated risk and mitigation, and management strategies are very broad themes in their realms. However, there are multifarious, imperceptive, and intricate aspects related to these themes that are impossible to cover in any single volume. This book aims to bring forth various case studies that enhance our understanding of the policy and planning processes to mitigate the losses done by natural hazards in the Himalayas by enhancing knowledge about the science behind it. In this chapter, we describe various issues and paradigm shifts within the field of hazards in the Himalayas and conclude with an outline of the chapters that form this book's basis.

Keywords Natural hazards · Landslides · Earthquakes · GLOFs · Floods · Himalayas

1 Introduction

Due to neo-tectonic morphology and dynamics, the natural hazards in the Himalayas are part and parcel of the communities living there. The Himalayas were always

S. Kanga · G. Meraj (✉)
Centre for Climate Change & Water Research (C3WR), Suresh Gyan Vihar University, Jaipur, Rajasthan 302017, India

G. Meraj · M. Farooq
Department of Ecology, Environment and Remote Sensing, Government of Jammu and Kashmir, Kashmir 190018, India

S. K. Singh
Centre for Sustainable Development, Suresh Gyan Vihar University, Jaipur, Rajasthan 302017, India

M. S. Nathawat
Department of Geography, Indira Gandhi National Open University, New Delhi, India

S. Kanga et al. (eds.), *Disaster Management in the Complex Himalayan Terrains*, Geography of the Physical Environment,
https://doi.org/10.1007/978-3-030-89308-8_1

sensitive to disasters as a region, but the unplanned and reckless development aggravated the situation and increased the risk of such disasters. As an impact on countries' socioeconomic development, such calamities adversely affect the already stressed economy (Banskota 2000; Meraj et al. 2018a, 2018b). Hence, the "development" that was intended to uplift the economy helps make it more stressed if done unplanned and by destroying vital natural supports. Intensive precipitation including cloudbursts, and earthquakes, augmented with anthropogenic mining activities and destabilization of slopes, are the main forces behind the causes of many major natural hazards in the Himalayas, such as floods, landslides, and earthquakes (Altaf et al. 2014; Joy et al. 2019; Carey et al. 2021). Moreover, climate change and unplanned development activities are some of the reasons behind the destruction caused by glacial lake outburst floods (GLOFs) in the Himalayas (Mishra et al. 2019). Every year in India and other Himalayan countries, extraordinary and intensive, longer-duration precipitation causes devastating floods (Ballesteros-Cánovas 2019). Although India has standard flood management protocols, the massive destruction it brings every year and the administrations' failure to restrict the loss of life and property implies a lack of planned strategy and policy failure (Islam et al. 2016).

Moreover, flash floods have become part and parcel of urban Himalayan commuters' daily lives in India, causing severe damage to their life and property (Dewan 2015; Kanga et al. 2017). Landslides, the movement of earth material moving down under gravity's influence, is often associated with earthquakes, intense rainfalls, and floods. Moreover, factors such as vegetation devoid mountain areas are particularly vulnerable to landslides, causing the river flow block (Nathawat et al. 2010; De Blasio 2011). If and when the river blocks explode, the settlements downstream could be devastated. Landslides in the hilly states of India are common. Himachal Pradesh, Uttarakhand, Shimla, and Kargil are some of the Indian Himalayan regions that witnessed floods due to the rivers' blockade by landslides in recent years.

The science behind the Himalayas' seismicity lies in the fact that since the Eocene period when India collided with Eurasia (some 45 million years ago), it is continuously migrating to the north that has resulted in the creation of the Himalayan mountain range, penetrating around 2000 km into Asia which continues till date (Aitchison et al. 2007). In the process, Himalayas are undergoing beneath the Eurasian plates, making them the most tectonically active mountains in the world. As a result, more than ninety percent of India's earthquakes are due to the seismicity of the Himalayas. Earthquakes in Uttaranchal (1991), Gujarat (2001), and Kashmir (2005) are some of the most catastrophic earthquakes in recent history. The damage in an earthquake event is mainly due to the unplanned development in seismic areas that cause widespread deaths and property loss (Gulati 2006). Moreover, glacial lake outburst floods (GLOFs) that occur when the moraine-dammed lakes at the high-altitude glacier sites break as a result of intense rainfall or earthquake, or any other factor has been one of the significant causes of damage in the Himalayan regions (Allen et al. 2016; Rafiq et al. 2019). The most havoc-wreaking was the GLOF event in the Chorabari Tal in the Kedarnath in June 2013, and the most recent event was in Uttarakhand in February 2021. Such events are one of the manifestations of anthropogenic climate change. The youngest mountain regions of the Himalayas are considered the most vulnerable parts of the world as far as forest fires are concerned (Thakur et al. 2020). Thousands of hectares of forests are destroyed by forest fires every year in India. This hazard is so alarming that India has set up a state-of-the-art forest fire alert system based on real-time satellite data and extensive ground-based monitoring and reporting incidents.

2 Concept and the Structure of This Book

South Asia harbors the complex Himalayan terrains with over one-fifth of the world's population and is recognized as the world's most hazard-prone region. The exponential increase in population with the consequent pressure on its natural resources and continued high rates of poverty and food insecurity makes this region the most vulnerable region to hazards in the world as far as the impacts of climate change are concerned. Over the last century, generally, the climatic trends in South Asia have been observed to be characterized by increasing air temperatures and an increasing trend in the intensity and frequency of extreme events. IPCC (2014) has reported that the Himalayan highlands shall face significant warming over the next century. Moreover, the disadvantaged people of this region's countries shall be more vulnerable due to the intense and frequent extreme weather events, such as heatwaves and severe precipitation episodes, as a result of the anticipated impacts of climate change. The increasing frequency of the natural hazards due to climate change impacts in the Himalayas calls for efficient management and policymaking in these regions, which the local governments can only implement through an established science-based robust action plan.

Disaster Management in the Complex Himalayan Terrains: Natural Hazard Management, Methodologies and Policy Implications is a comprehensive edited book focusing on the science and management aspect of natural hazards using innovative spatial information sciences, satellite remote sensing, and mathematical modeling. The book is aimed to replenish the gap in the available literature by bringing the concepts, theories, and practical experiences of the specialists and professionals in this field together in one volume. There is a shortage of such a book; hence, the editors have tried hard to bring the best literature in this field in the form of the book to help students, researchers, and policymakers develop a complete understanding of the management and policy implications of natural hazards in the Himalayan region.

This volume contains effective methods, science, and management that form the title of the book. The book is divided into five parts. The first part (Part I) includes this chapter and Chap. 2 and focuses on the introduction to the disasters and hazards in the Himalayan terrains. Chapter 2 elaborately discussed the damage witnessed due to massive hazards to natural and man-made environments in South Asia by reviewing the latest literature in the field. The chapter focused on floods, GLOFs, and landslides and argued that understanding the processes and driving forces for different types of hazards is a prerequisite for properly forecasting and mitigating hazards in the Himalayas.

Part II discusses the causes and consequences of landslides in the Himalayas and contains Chaps. 3–5. According to Chap. 3, the Himalayas are characterized by the highest cases of landslides due to fragile rocks and major tectonic boundaries, resulting in the highest intensity and frequency of earthquakes. The authors focused on various modern techniques for systematic studies highlighting the landslide's extent and effect to suggest proper remedial measures. Chapter 4 focuses on the landslide causes and genesis and mitigation measures of the NH-44A Srinagar Jammu Highway in India. The authors argue that unless proper scientific procedures are formulated, the condition of the Kashmir valley's lifeline road is not going to get better soon. Chapter 5 discussed the geoenvironmental impacts of the road widening project of the NH-44A highway that has a complex ecosystem and rugged topography. The authors argue that climatic and hydrometeorological factors are playing as a catalyst in construction complications. Implementation of stepwise preventive measures can only reduce the risk of geoenvironmental hazards, particularly landslides in the area. The author proposes a cladding wall, retaining wall, micro-piling at the tower foundation with a waler beam, anchor, and shotcrete to minimize the consequences of any possible landslides. It is suggested that an independent Hill Development Authority may be constituted, mandated for preparing long-term plans for hill development in this part of Himalaya after taking care of local geological challenges. The study argues that landslide study during the project feasibility stage would save the project completion time and project cost escalation.

Part III discusses the causes and consequences of floods in the Himalayas and comprises Chaps. 6–8. Chapter 6 discussed the September 2014 floods in Jammu and Kashmir's union territory and their causes, impacts, and mitigation strategies. Extreme rainfall throughout the Kashmir valley during the first week of September 2014 has been assumed to have triggered these floods augmented by the rapid snowmelt. The authors suggested long-term mitigation measures by developing an efficient early warning system and enhancing the watershed management practices using the state-of-the-art remote sensing and GIS techniques. Chapter 7 discussed different types of floods in the Himalayas and remote sensing data to estimate river and glacial lake bathymetry. The authors specifically discussed various methods to model and map glacier lake outburst floods (GLOFs), which are considered devastating natural hazards in the Himalayan region. Chapter 8 discusses a huge shortage of knowledge regarding the characterization and vulnerability of energy resources such as small-scale hydropower to extreme events, land-use changes, and their social contestations in the watersheds. The study carried out in the Gagas watershed (Western Himalaya) focused on watermills' sustainability as small-scale hydropower using field, satellite remote sensing, and geographic information system (GIS). Around 12% of watermills were observed to be functional, and 88% were non-functional. The study argues that the frequent occurrences of extreme events such as floods, with 31% of the total area within very high to high flood susceptibility, aggravate the expenditures on maintenance of small hydel projects, thus impacting the local economy.

Part IV discusses the assessment and forecasting of earthquakes in the Himalayas and comprises Chaps. 9 and 10. Chapter 9 monitored earthquake-related deformation using satellite technology, known as Interferometry Synthetic Aperture Radar (InSAR), to derive deformation rates to a moderate earthquake that occurred in the western part of Nepal near Dipayal Silgadhi on November 19, 2019. The authors used 42 interferograms covering the epicentral area with latitude ranging from 29.1°N to 29.8°N and longitude ranging from 80.5°E to 81.8°E. Such studies are critical due to the role they can provide in the mitigation and management of earthquake-related losses in the complex Himalayas. Further, Chap. 10 has discussed the implementation of a neural network-based earthquake forecasting model that involves deep learning algorithms to detect and locate earthquakes in an effective way. The authors considered eight seismicity indicators to form the input of the proposed neural network and observed that the proposed network provides 90% accuracy and an F1 score of 0.89 for the earthquake data during 1980–2020. Such results undoubtedly are extremely to develop new standards of earthquake monitoring and mitigation measures in the Himalayan subcontinent.

Part V discusses the hazard mitigation strategies in the Himalayas and comprises Chaps. 11 and 12. Chapter 11 discussed India's forest fire alert system with a special reference to the UT of Jammu and Kashmir's fire vulnerability assessment. Using actual forest fire incidences from Jammu and Kashmir forest department for 2002–2018, and MODIS Satellite Fire Data (2012–2018), forest division, ranges, and compartment boundaries forest fire vulnerability assessment of the UT of Jammu and Kashmir has been carried out. A correlation index of MODIS (3141) and actual fire incidences (2438) from 2012 to 2018 was derived, which showed a correlation coefficient of 0.97 using the seven-year dataset of actual fire points. Since the points derived from MODIS satellite data cover a larger extent, the number of points varies with actual forest fire points. The results showed many regions of the UT fall in the high fire vulnerability category. They can only be managed if the forest fire alert system of the UT is followed very strictly. Chapter 12 elaborately discussed the impact of climate change on the patterns of precipitation across the world. As a result, there is an increase in the disasters like flash floods, glacier lake outburst floods (GLOFs), snow avalanches, landslides, landslide lake outburst floods (LLOFs), etc. The authors showed that extreme precipitation events that increased over India in the last 113 years have resulted in the region's surge of disasters. The authors have shown how established scientific literature has shown that warming due to climate change has dire consequences over the Himalayas. At the end, the authors discussed some of the policy and decision-making strategies taken by the Himalayan region of Kashmir.

3 The Way Forward for Disaster Risk Reduction in the Himalayan Region

A diverse range of different plant species can be found in the Himalayas covering almost four million square kilometers (1.6 million square miles) from Afghanistan to Myanmar. One of the most biodiversity-rich places is also the dangerous. Since the terrain is so steep and subject to earthquake and landslide activity and heavy rainfall and snowfall, the Himalayas are especially susceptible to floods, landslides, avalanches, and earthquakes. It has been discussed throughout this book how, to date, the number and severity of natural hazards within the Himalayas are on the rise, due in part to climate change. A general threat to life and livelihoods is posed by environmental degradation. Though a community's vulnerability to natural hazards also includes exposure to disasters and their impact, they can also result in property and personal injury. While there are some aspects of risk and exposure that stem from the physical and environmental aspects, socioeconomic factors such as poverty, human settlement, habitat, lack of preparedness, adaptive capacity, and susceptibility all impact how vulnerable the population is.

While many families from the region struggle to rebuild their homes and livelihoods, poverty leaves them with few resources. A widening gap between the rich and the poor exists in the region. These facts point to an increased risk of natural disasters (see Chaps. 2 and 5). Hazard, exposure, and vulnerability are the three key factors that drive disaster risk. Changes in the environment, such as global warming, are causing natural hazards to grow in both intensity and occurrence. The number of people and the value of property exposed to a hazard and the community's overall vulnerability all increase its exposure to that hazard. The IPCC's latest assessment report further expands on the interactions among these critical elements for handling climate change risk. Chapters 11 and 12 adopt a policy framework to reduce risk and increase resilience due to this understanding of disaster risk, hazard, exposure, and vulnerability. There are some critical areas that policy and decision makers must set up to cope with natural hazards. These include multi-hazard environment assessment, assessing the impacts of climate change and variability, managing accessibility and connectivity in the most vulnerable regions, and formulating regional governing bodies that could take the appropriate steps at the times of emergencies.

Enhancing community resilience to hazards by lowering vulnerability and pursuing resilience building measures necessitates a thorough understanding of catastrophe risks, which can aid policymakers in prioritizing actions that will raise their population's resilience to catastrophic disasters. A framework is needed to assess risks from hazard events and recommend measures to help communities in the Himalayas become more resilient. It includes some of the most basic disaster risk reduction elements such as planning and execution, command-and-control mechanisms, monetary incentives, persuasion through knowledge, and early warning systems. Hence, depending on the type of hazard event under consideration for resilience building, a comprehensive program taking into cognizance the science behind the evolution of such a hazard will pave way for the effective and long-term mitigation strategy.

References

Aitchison JC, Ali JR, Davis AM (2007) When and where did India and Asia collide? J Geophys Res Solid Earth 112(B5)

Allen SK, Rastner P, Arora M, Huggel C, Stoffel M (2016) Lake outburst and debris flow disaster at Kedarnath, June 2013: hydrometeorological triggering and topographic predisposition. Landslides 13(6):1479–1491

Altaf S, Gowhar M, Romshoo SA (2014) Morphometry and land cover based multi-criteria analysis for assessing the soil erosion susceptibility of the western Himalayan watershed. Environ Monitor Assess 186(12):8391–8412

Ballesteros-Cánovas JA, Allen S, Stoffel M (2019) The importance of robust baseline data on past flood events for regional risk assessment: a study case from the Indian Himalayas. UNISDR Global Assessment Report

Banskota M (2000) The Hindu Kush-Himalayas: searching for viable socioeconomic and environmental options. Growth, poverty alleviation, and sustainable resource management in the mountain areas of South Asia, 57–106

Carey M, McDowell G, Huggel C, Marshall B, Moulton H, Portocarrero C, Provant Z, Reynolds JM, Vicuña L (2021) A socio-cryospheric systems approach to glacier hazards, glacier runoff variability, and climate change. In: Snow and ice-related hazards, risks, and disasters. Elsevier, pp 215–257

De Blasio FV (2011) Introduction to the physics of landslides: lecture notes on the dynamics of mass wasting. Springer Science & Business Media

Dewan TH (2015) Societal impacts and vulnerability to floods in Bangladesh and Nepal. Weather Clim Extremes 7:36–42

Gulati B (2006) Earthquake risk assessment of buildings: applicability of HAZUS in Dehradun, India. Enschede, ITC

Islam R, Kamaruddin R, Ahmad SA, Jan SJ, Anuar AR (2016) A review on mechanism of flood disaster management in Asian. Int Rev Manage Market 6(1)

Joy J, Shruti K, Singh SK (2019) Kerala flood 2018: flood mapping by participatory GIS approach, Meloor Panchayat. Int J Emerg Techn 10(1):197–205

Kanga S, Singh SK, Sudhanshu. (2017) Delineation of urban built-up and change detection analysis using multi-temporal satellite images. Int J Recent Res Aspects 4(3):1–9

Meraj G et al (2018b) An integrated geoinformatics and hydrological modelling-based approach for effective flood management in the Jhelum Basin, NW Himalaya." Multidisciplinary Digital Publishing Institute Proceedings 7.1: 8

Meraj G et al (2018a) Geoinformatics based approach for estimating the sediment yield of the mountainous watersheds in Kashmir Himalaya, India. Geocarto Int 33(10):1114–1138

Mishra A, Appadurai AN, Choudhury D, Regmi BR, Kelkar U, Alam M, ... Sharma U (2019) Adaptation to climate change in the Hindu Kush Himalaya: Stronger action urgently needed. In: The Hindu Kush Himalaya assessment. Springer, Cham, pp 457–490

Nathawat MS et al (2010) Monitoring & analysis of wastelands and its dynamics using multiresolution and temporal satellite data in part of Indian state of Bihar. Int J Geomat Geosci 1(3):297–307

Rafiq M, Romshoo SA, Mishra AK, Jalal F (2019) Modelling Chorabari Lake outburst flood, Kedarnath. India. J Mountain Sci 16(1):64–76

Thakur S, Negi VS, Pathak R, Dhyani R, Durgapal K, Rawal RS (2020) Indicator based integrated vulnerability assessment of community forests in Indian west Himalaya. Forest Ecol Manage 457:117674

Hazards in the Perspective of Himalayan Terrain: A Review

Fayma Mushtaq, Afaan Gulzar Mantoo, Anamika Shalini Tirkey, and Sofi Zubair Ahmad

Abstract The unique geological and geographical setting of the Himalayan region makes it highly susceptible to various types of natural as well as anthropogenic hazards causing destruction and leading to loss of valuable human life and precious resources. The region has witnessed massive hazards as well as disasters over the years that have caused severe damage to natural as well as man-made environments. Degrading environment, immensely increasing population, unscientific and unplanned constructional patterns, and poor mitigation strategies increases the risk of the hazards in the region by many folds. This chapter makes an attempt to review the occurrences of various hazards in the Himalayan region along with the recorded and studied events. The major and prime focus of this chapter runs mainly around Floods, GLOFs, and Landslides. It is argued that understanding the processes and driving forces for different types of hazards is a prerequisite for proper forecasting and mitigation. The changing climate and landscape of the region demands for better monitoring, mitigation strategies, capacity building, and socioeconomic empowerment for reducing the scale of damage caused by these hazards.

Keywords Natural hazards · GLOFs · Landslides · Floods · Remote sensing and GIS · Himalayas

F. Mushtaq (✉) · A. G. Mantoo · S. Z. Ahmad
Department of Ecology, Environment and Remote Sensing, Government of Jammu and Kashmir, Srinagar 190018, India

A. S. Tirkey
Department of Geoinformatics, School of Natural Resources and Management, Central University of Jharkhand, Brambe, Ranchi, Jharkhand 835205, India

S. Z. Ahmad
Institute of Tibetan Plateau Research, Chinese Academy of Sciences ITPCAS, Beijing, China

S. Kanga et al. (eds.), *Disaster Management in the Complex Himalayan Terrains*, Geography of the Physical Environment,
https://doi.org/10.1007/978-3-030-89308-8_2

1 Introduction

The word hazard in English refers to "Chance of harm or loss, risk" (Ghimire 2011). At the commencement of the term, "hazard" was recognized as "natural hazards" stated as those extraneous forces of the physical environment which are harmful to man (Bertram et al. 1979).

United Nations Office for Disaster Risk Reduction (UNDRR) defines hazard as "any grave disturbance in a community due to dangerous events that interact with exposure conditions, vulnerability and capacity, resulting in one or more of the following: human losses and impacts on economy and environment. United Nations Disaster Relief Organization (UNDRO) defined it as the probability of occurrence, within a specific period in each area, which is a potentially damaging phenomenon (Shi 2019). According to United Nations International Strategy for Disaster Reduction (UNISDR 2009), any natural process that can pose potential negative impact on the society, ecology or the economy can be termed as a "Hazard." The most important characteristics of any hazard include its magnitude, intensity, and frequency of occurrence of events (Ali et al. 2016).

Broadly, Hazard can be classified into two categories as natural and anthropogenic hazards. Integrated Research on Disaster Risk (IRDR 2014) classifies hazards into 6 different families, 20 main events, and 47 perils. The six broad categories of hazards within the family group are: Geophysical hazard, Hydrological hazard, Meteorological hazard, Climatological hazard, Biological hazard, and Extraterrestrial hazard (IRDR 2014). Natural hazards have events associated with them and can be classified broadly as geophysical (volcanic activities, earthquake, landslides), hydrometeorological (tropical storms, drought, floods), and biological (pandemic, epidemic). Anthropogenic hazards include human induced events such as environmental degradation, climate change, mining of non-renewable resources, and scientific hazards (UNISDR 2009).

Hydrometeorological disasters are the atmospheric and hydrologic processes that may cause significant damage to life, infrastructure, economy, and environment (Ali et al. 2016). Hydrometeorological hazards are caused by extreme meteorological and climate events, such as tropical cyclones thunderstorms, floods, hailstorms, heat waves, droughts, heavy snowfall, hurricanes, avalanches, tornadoes, coastal storm surges, landslides (Wisner et al. 2014; Wu et al. 2016). Human and economic losses to natural hazards have escalated in recent decades (Bouwer 2011). Among the major disasters in the world, hydrometeorological disasters account for over one-third of damage including loss of life, economy, and infrastructure (Jayawardena 2015) (Fig. 1).

The Himalayas are the mountain ranges spanning 2400 km that separate the Tibetan plateau from the Indian subcontinent. The Himalayas are stretched over 5 countries namely, China, India, Pakistan, Bhutan, and Nepal, and are home to about 53 million people (Apollo 2017). Being young mountains, the Himalayas have very complex topography which makes them more vulnerable to different types of hazards, especially the hydrometeorological ones. The region is highly flood-prone

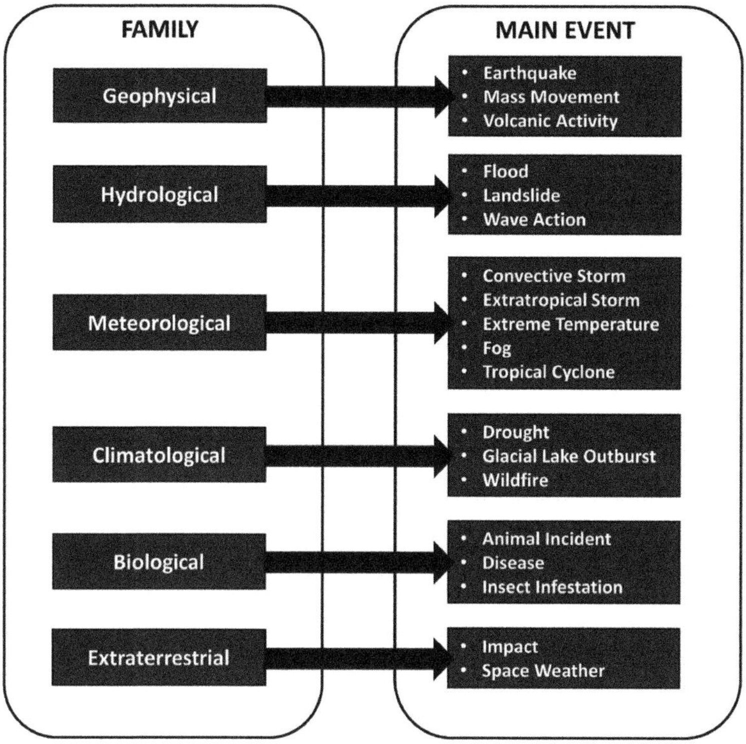

Fig. 1 Classification of natural hazards into six broad categories (IRDR 2014)

because of enhanced and varying orographic precipitation trends with steep slopes having very thin soil cover over impervious substratum or bedrock (Arora et al. 2016). This results in higher runoff coefficients alongside limited storage capacity in mountainous terrain causing severe flash floods accompanied with debris flow. Moreover, the precipitation regimes over the Himalayas and its constituent river basins are highly variable. The annual mean precipitation over the Brahmaputra, the Ganges, and the Indus river basins is estimated as 2143, 1094, and 435 mm, respectively (Nepal and Shrestha 2015). This variability is due to the presence of different climatic disturbances including the Western disturbances or Westerlies (Dimri et al. 2015; Madhura et al. 2014) and the Monsoons (Shrestha 2008; Singh and Ranade 2010).

Hazard risk studies often stress on the impacts of an individual hazard, such as landslides (Althuwaynee et al. 2014; Pellicani et al. 2017), floods (Kabenge et al. 2017; Kazakis et al. 2015), droughts (Lehner et al. 2006), and earthquakes (Dhar et al. 2017; Theilen-Willige 2010). The Himalayan region is prone to multiple hazards that could occur concurrently or manifest in a series of cascading events (Kappes et al. 2012). This chapter is an attempt to review and understand the studies that focus around understanding, assessing, and mitigating strategies different hydrometeorological hazard in the Indian Himalayan region. In terms of rainfall pattern, drainage networks, geology, land use/land cover, and lithology, the Himalayan region is one of the most diverse and heterogeneous area of the world (Dikshit et al. 2020). Extending from 26° 20′ to 35° 40′ N and 74° 50′ to 95° 45′ E, the region covers around 500,000 km of India's total landmass (Ministry of Environment, Forest and Climate Change, Government of India). Many states and Union territories or some part of them are located in this region including Arunachal Pradesh, Assam, Himachal Pradesh, Jammu and Kashmir, Ladakh, Manipur, Meghalaya, Mizoram, Nagaland, Sikkim, Uttarakhand, and West Bengal.

1.1 Floods

Floods are the third most devastating hydrometeorological disasters, when considering the number of deaths recorded (Komolafe et al. 2019). It is estimated to have generated more than 30% of all disasters globally during 1945–1986 (Glickman 1992) with a high rate of occurrence affecting many countries worldwide and causing huge economic and human loss annually (Mohapatra and Singh 2003; Thilagavathi et al. 2011; Wilby and Keenan 2012). The populations and infrastructures are at continuous risk to flooding owing to increased water cycle worldwide and accelerated population growth (Webster et al. 2005). The economic activities and populations in floodplains are growing rapidly globally which has increased their socioeconomic exposure to floods (Hirabayashi et al. 2013; Winsemius et al. 2016). It is mainly caused by uneven distribution and high-intensity rainfall (Dhar and Nandargi 2003; Sanyal and Lu 2005). Floods are the result of intricate hydrological, geological, and geomorphological conditions, deforestation, and urbanization that cause significant social, economic, and environmental damage (Mukherjee and Singh 2020; Skilodimou et al. 2019). The damage caused by flooding depends mainly on the duration and intensity of precipitation, the soil type, slope, and land use/land cover (Lal et al. 2020). Apart from these, climate change is also influencing the frequency of floods and damages associated with it (Gupta and Nair 2011; Milly et al. 2002), wherein the last three decades has witnessed increased flood occurrences (Rozalis et al. 2010). The present situation is now more worsened owing to changes in amount, seasonality, and event-size distribution of rain events. The percentage of flood occurrences to global natural disasters during 1985–2009 accounts to be 40% of the total natural disasters, and therefore there is excessive destruction in terms of economic and human losses (Komolafe et al. 2018; Ferreira et al. 2011). The impact of floods is evident from suffering populations, damage to infrastructure, damage to crops and animals, loss of ecosystem services, the spread of diseases, and contamination of water supply (Komolafe et al. 2020).

Globally, the frequency of natural disasters, such as floods, hurricanes, and droughts, has grown from an average of 50+ incidents in the 1970s to 350+ events in 2014 and 500+ events in 2000. Between 1970 and 2006, the number of floods increased from fewer than 50 to 226, and between 2007 and 2008, the number of floods increased to 218. The count of storms has gone from under 50 in the 1970s to many more than 120 in the 1990s and the latter part of the last decade. It hovered between 50 and 60 in every year between 2000 and 2010. Additionally, there has been an increase in the frequency of droughts, but not in the intensity or spread area. (DTE). Flood disasters resulting from extreme rainfall have increased in recent years in many parts of the world (Jayawardena 2015). India has also witnessed extreme rainfall events leading to flood, and huge loss of lives (Guhathakurta et al. 2011). The mountainous regions in India particularly the Himalayas are prone to water-related disasters. The Himalayas and adjacent low-lying areas seldom have flood instances owing to extreme precipitation due to synoptic climatic, artificial, and natural dam bursts. The runoff in the Himalayan basins is sensitive to glacial melt (Singh and Bengtsson 2004; Ma et al. 2010; Jeelani et al. 2012) and therefore the situation becomes all the more devastating when heavy rain fall events are augmented with high runoff volumes resulting from melting snowpack in the Himalayas. Some recent studies like the study of Houze et al. (2011) provide synoptic analysis of floods in the Himalayas.

The flood mitigation measures in developed countries, incorporates structural means which maybe unaffordable in most developing countries. Therefore, early warning systems are being adopted in conjunction with non-structural means, which are cost-effective and sometimes the only option. The early warning system is much helpful in protecting human lives and minimizes the damages from a disaster that are above a certain critical level. An early warning system comprises of many inter-related components like, forecasting, transformation of the forecast into warning, the transmission of warning to local decision-makers, and conversion of the warning into remedial action. To forecast an impending event, it is necessary to understand causes and effects in quantitative terms and to formulate a mathematical model that links cause and effect. A list of flood modelling techniques as an approach to flood early warning systems is given in Fig. 2 (Nkwunonwo et al. 2020).

Satellite remote sensing offers new possibilities for flood estimation from regional to global level and provides information about changes in surface water dynamics through direct observations using optical remote sensing (Brakenridge and Anderson 2006; Ordoyne and Friedl 2008) or Synthetic Aperture Radar (SAR) (Horritt et al. 2003; Mason et al. 2012) and hydrological modeling using remote sensing inputs of precipitation, land cover, vegetation, topography, and hydrography (Shrestha 2008; Wu et al. 2012). One of the main challenges of researchers working on flood monitoring and flood forecasting are the lack of availability of ground truth data for validating and quantifying the uncertainty of historic and ongoing flood events, for example, the Dartmouth Flood Observatory (DFO) (http://floodobservatory.colorado.edu/) (Brakenridge and Anderson 2006), NASA Near Real-Time Global MODIS Flood Mapping (http://oas.gsfc.nasa.gov/floodmap/), the Global Flood Monitoring System (GFMS, (http://flood.umd.edu/), (Wu et al. 2014), and Global Flood Awareness System (GloFAS) (http://www.globalfloods.eu/) (Alfieri et al. 2013). The Acoustic Doppler Current Profiler (ADCP) onboard remote-controlled boats are extremely important for obtaining flood information like water

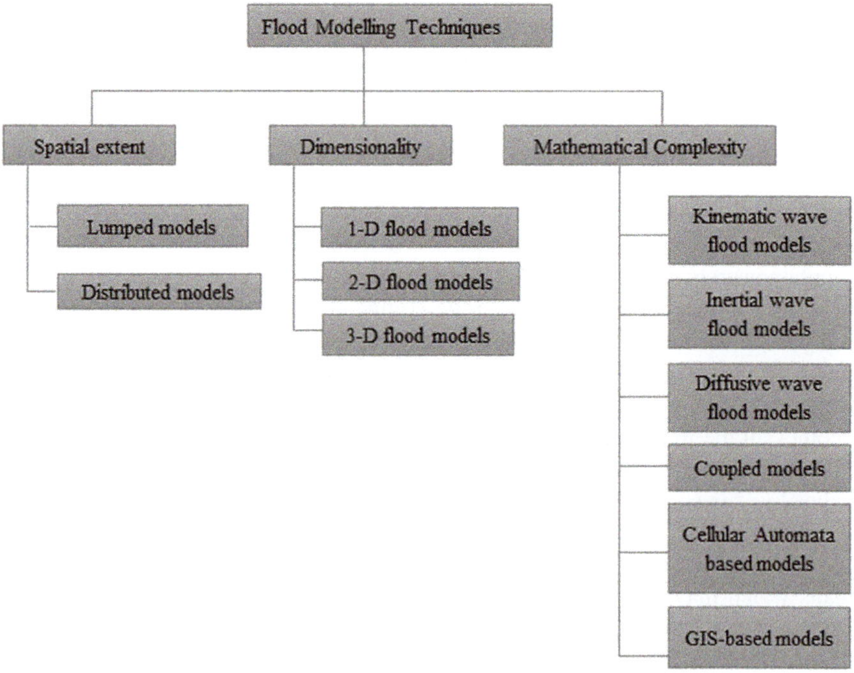

Fig. 2 Classification of various flood modeling techniques (Nkwunonwo et al. 2020)

discharge and velocity profile for ongoing flood events and has proved to be an important complementary tool to satellite and air-based sensors. Figure 3 shows the Global Flood Monitoring System (GFMS) for a widespread flood event in south China in July 2016. The European Space Agency (ESA) through Sentinel satellite data now provides real-time radar-based satellite imagery for flood hazard mapping (Kaku and Held 2013; Kwak 2017; Twele et al. 2016).

The existing flood modeling techniques can be classified into one-dimensional, two-dimensional and three-dimensional models (Bates et al. 2005; Ervine and

Fig. 3 GFMS-based flood inundation map for Yangtze River basin (Wu et al. 2016)

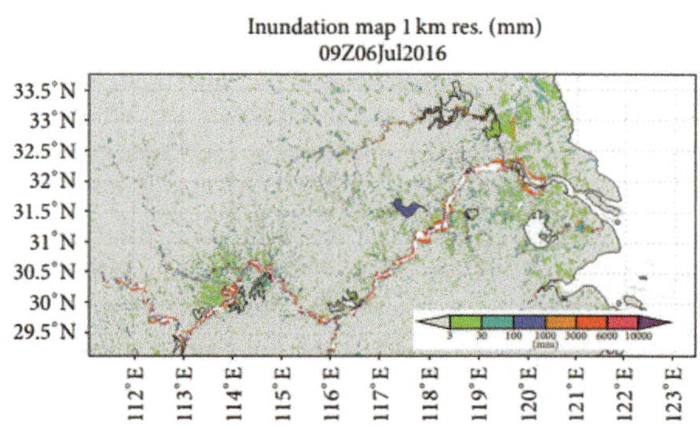

MacLeod 1999; Haider et al. 2003). The ISIS, MIKE 11, and HEC-RAS are one-dimensional flood models that represent the channel and floodplain as a series of cross-sections perpendicular to the direction of flow and solve the full approximation or some of the shallow water one-dimensional equations (SWEs) (Bates and De Roo 2000). The one-dimensional flood models are the simplest of all flood models, as they are computationally efficient, and are convenient for parameterization using traditional field surveying, without the distribute topographic and friction data (Bates and De Roo 2000).

Two-dimensional flood models such as TUTFLOW, SOBEK, and MIKE 21 are also being applied progressively in flood modeling as they solve the two-dimensional SWEs using appropriate numerical schemes (Dottori and Todini 2013; Abderrezzak et al. 2009; Mignot et al. 2006; Soares-Frazão et al. 2008). Advances in remote sensing technology (particularly through high-resolution and highly accurate input data like LiDAR and SAR data along with and improved computing capacity have increased the popularity of two-dimensional models (Hunter et al. 2008). The three-dimensional flood model considers the flow of floodwater to be completely three-dimensional and therefore solves the Navier-Stoke equations (Néelz et al. 2009). In fact, in order to dynamically represent the physics of water flow, especially in the urban areas of developing countries, it is worth applying the three-dimensional model. A vital factor that should not be overlooked when reviewing flood models from a dimensionality perspective are the schematics or numerical formulations that are critical to enhance the scope of flood modeling functionality (Szewrański et al. 2018; Ward et al. 2015). In practical application, these formulations are often evaluated by discretization of a meshed topographic surface. This highlights the importance of quality satellite datasets in flood modeling and the possible implications for data poor areas. Various numerical schemes have been formulated over the years, to solve a wide variety of hydrodynamic problems, particularly in computational mathematics and flood modeling (Casulli and Cheng 1992; LeVeque 1997). The characteristics schemes, explicit and implicit finite difference schemes, semi-implicit finite difference schemes, finite elements, and finite volume numerical schemes are some of the widely applied numerical schemes by various researchers (Bradford and Sanders 2002; Dumbser et al. 2015; George 2011; Abderrezzak et al. 2009). The simplified two-dimensional models or reduced complexity models (RCM) such as LISFLOOD-FP, JFLOW, and ISIS-FAST are some of the mathematically simple flood modeling techniques (Bates and De Roo 2000; Yu and Lane 2006). The RCM solves the kinematic wave, diffusive wave, and inertial wave equations resulting from various simplifications of the SWEs in conjunction with other raster-based flood models (coupled 1D/2D models), and Cellular Automata (CA)-based models (Hunter et al. 2007; Patro et al. 2009). So far, these models have provided realistic applications in urban flood modeling, although they still pose critical problems in urban flood modeling within the developing countries (Liu et al. 2015). The CA-based flood models in particular have recently become increasingly important (Dottori and Todini 2011). To simulate a flood, CA-based flood models use transition rules in a discrete space within specific neighborhoods (Ghimire et al. 2013; Rinaldi et al. 2007).

1.2 Landslides

Mountainous regions across the globe have been affected by landslides including Himalayas. Most of the studies carried out have been focused around major landslides which involve traditional field-based work (Turner and Schuster 1996). These studies provide an insight into the landslide damage assessment and possible driving factors behind the events, but fail to define the relationship with the morphometry. Identification and visualization coupled with standardized classification are basics for pre-/post-hazard analysis of landslides and related events that can help in landslide zonation, rescue and relief operations, and mitigation strategies (Martha et al. 2010). Various image classification techniques have been developed to map and analyze landslides which can be credited to the development of various remote sensing technologies, availability of high-resolution data (spatial and temporal), and researchers. With the use of high-quality satellite images researchers can identify the changes in soil structure and vegetation loss post landslide event (Martha et al. 2012). Object-based (OB) and Pixel-based (PB) categorization are two basic procedures used for landslide damage assessment using high and medium resolution imagery, respectively, (Dikshit et al. 2020). In the Indian Himalayan region, the object-based image classification technique has been primarily used. Another approach mostly used is the time-series analysis of land use/land cover change (Martha et al. 2016) (Fig. 4).

Object-based (OB) techniques have been extensively used in landslide detection after a landslide event has occurred, as well as for creating a landslide inventory from historical data. (Martha et al. 2010) used LISS-IV data (5.8 m) and 10-m Cartosat-1 derived DEM to prepare a detailed landslide inventory Okhimath region in Uttarakhand for the year 1998. The accuracy for recognition and classification of

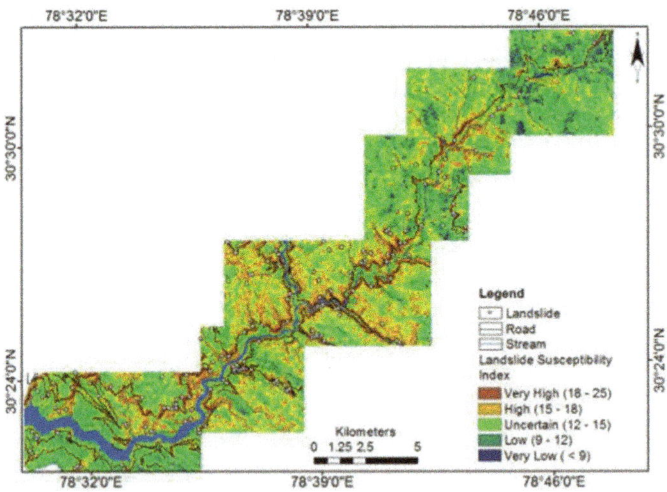

Fig. 4 Tipari to Ghuttu highway corridor landslide susceptibility zonation (Pandey and Sharma 2017)

number of landslides using semi-automatic approach in the region was determined to be 76% and 69%, respectively. Subsequently, (Martha et al. 2012) used panchromatic images (from 1998 to 2006) acquired from IRS-1D (5.8 m) and Cartosat-1 (2.5 m) for the same region to prepare a landslide inventory database. The study depicted the importance of texture in case of missing spectral information.

Martha et al. (2013) analyzed time-series satellite data of Resourcesat-1 (5.8 m), IRS-1D panchromatic (5.8 m), and Cartosat-1 (2.5 m) data from 1997 to 2009 in Okhimath region of Uttarakhand. Wherein it was found that the accuracy of the landslide detection ranged between 60 and 89% and for landslide extent it varied between 71 and 97%. (Ghosh et al. 2020) prepared landslide inventory in GIS environment of the Mandakini valley of Uttarakhand Himalayas for the period 1962–2013 incorporating both conventional data sources and different Earth Observation (EO) data. A thorough landslide inventory was compiled during the study which could be used to study to understand the mechanism of landslides and related events. It could be used to understand possible triggering factors, susceptibility mapping and morphometric conditions that determine the scale of the event (Fig. 5).

The identification and classification of landslide events have also been carried out in the Himalayan region of Uttarakhand using very high-resolution satellite data including Resourcesat-2 (5.8 m), Kompsat-2 (1 m), Cartosat-2 (1 m), GeoEye-1 (0.5 m), and digital elevation data such as Cartosat-1 DEM (10 m) (Martha and Kumar 2013). The comparison of pre and post-disaster images has been carried out by (Martha et al. 2015) to identify the June 2013 Uttarakhand event using Resourcecat-2, GeoEye-1, and Cartosat-2a. (Martha et al. 2016) also identified new landslide events based on a comparison of pre and post-disaster images. Vamsee et al. 2018 used Resourcesat-2 images in optimal scale parameter selector (OSPS) tool in Uttarakhand to improve the scale element of the MRS technique. The use of freely available

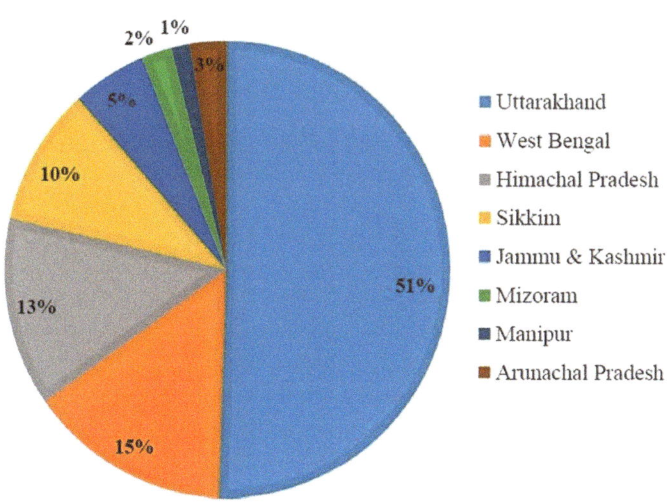

Fig. 5 Landslide studies in the Indian Himalayan region (Dikshit et al. 2020)

Google Earth (GE) has been incorporated in landslide mapping by (Kumar et al. 2019) along Satluj Valley in Northwest Himalaya. (Kumar et al. 2018) used Google Earth images of 2005, 2012 and 2014 to analyze changes in Pawari landslide.

Landslide hazard assessment is carrying out using landslide vulnerability mapping which considers different influencing factors and their spatial distribution to determine the potential of landslide event (Reichenbach et al. 2018). The techniques used for susceptibility mapping include Qualitative, Semi-quantitative, Quantitative, and Deterministic (Dikshit et al. 2020). The qualitative methods are quite subjective where values are determined based on the knowledge about the study area (Anbalagan 1992). Semi-quantitative methods are primarily based on different logical tools and stress on the significant factors by assigning weights like weighted linear combination, danger pixel approach and Analytic Hierarchy Process (AHP) (Kanungo et al. 2008). Also, the quantitative models can be categorized as bivariate as well as multivariate which is directly dependent on the landslide density under every inducing factor (Kumar and Anbalagan 2015). Bivariate as well as multivariate statistical models are used to compute weights, but multivariate models are dependent on collective effect of all the parameters (Ghosh et al. 2011; Mandal and Mandal 2018). Logistic regression is the most popular multivariate model used (Das et al. 2012). Bivariate statistical models include information value, frequency ratio, and fuzzy logic methods (Mathew et al. 2007; Pourghasemi et al. 2018; Sharma and Mahajan 2019). TRIGRS, SINMAP, SHALSTAB, SHETRAN, etc., are some of the many deterministic models that are used to analyze the slope stability. The analysis of mechanical and physical properties of soil computed in form of a factor of safety (FS) determines its susceptibility to landslides. (Sarkar et al. 2016) used SHALSTAB to determine the critical steady-state rainfall for slope stability in Darjeeling region of Indian Himalayas.

Computational techniques such as machine learning (ML), artificial neural network (ANN) and support vector machine (SVM) have efficiently proved to outperform traditional approaches (Peethambaran et al. 2019; Pham et al. 2017, 2019, 2020). (Ramakrishnan et al. 2013) used backpropagation neural network and determined a prediction capability of 80%. Pham et al. (2017) used different machine learning (ML) models based on SMO-SVM, VFI, LR, FT, MLP-neural networks, and NB, that depicted better capability for an area of 323 km^2 in Uttarakhand region. Pham et al. (2019, 2020) used four hybrid machine learning (ML) models for Pittorgarh region (242 km^2) in Uttarakhand and were found quite effective (Fig. 6).

The landslide inventories of the area show that most of the landslides have occurred along the highway corridors, particularly in the Lesser Himalayan rock formation which is highly weathered and fractured such as phyllite and schist-quartzites. It is concluded that the majority of the slope failures were triggered by rainfall in the area were debris fall, debris flow, and slump. Even a moderate-intensity rainfall episode has been seen to cause shallow landslides along the stream, slope cutting, and excavation surface along the highway corridors. Two main situations were identified: first, high-intensity rainfall episodes which increase incumbent load due to infiltration, give rise to numerous slides and falls; and second, anthropogenic activities such as road construction led slope cutting and excavation creates vulnerable zones

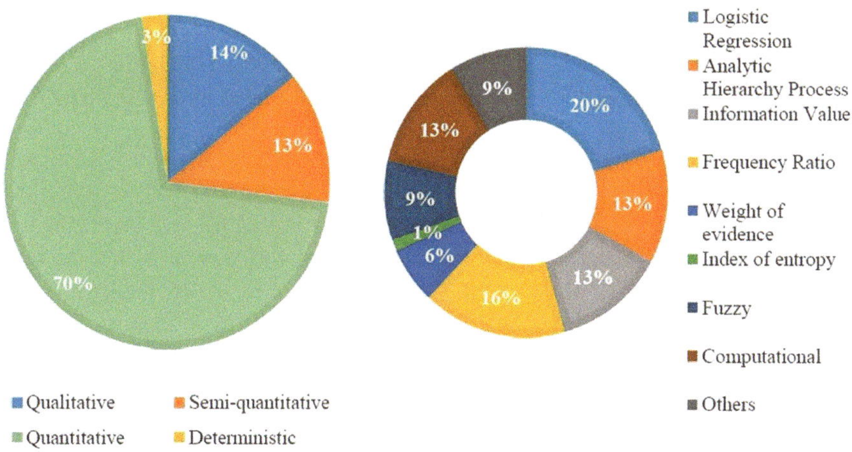

Fig. 6 Types of models and techniques used in hazard susceptibility studies (Dikshit et al. 2020)

and land use, which causes larger landslides, rotational and complex slides in the Lesser Himalayan metasedimentary rocks such as schist, phyllite, and conglomerate. However, apart from the triggering factors (rainfall and anthropogenic activities); slope inclination, curvature, and land use were also seen to influence the landslide distribution. The cluster of landslides was mapped on the topographies with slope angles ranging 29°–43°, inclined in the southern direction, particularly in the region of phyllite and schist-slate lithology. These areas are covered by barren land and scrubs land with NDVI values less than 0.35. These slope failures can be classified under debris fall and slide. Debris flows which were largely noted along the streams and circumference of the reservoir showed slope angles ranging from 21° to 33°. The landslide susceptibility zones prepared in the study area using the FR method shows high LSI along the highway corridors, streams, barren land, and scrubland areas, which are 15 highly weathered lithologies, southerly inclined slope direction. These areas having LSI values greater than 18 show frequent landslides in the area. The validation statistics of the FR model tested using 50 random landslide locations reveal 81.2% predictive accuracy of the model. This method is easy to compute has produced highly accurate landslide prediction susceptibility in the study area. For any landslide event, we can easily predict the transition from a stable creep to catastrophic landmass failure. The widespread improvements in topographic resolution and prediction of precipitation coupled with landslide modeling can help in minimizing the time response in case a landslide event happens to occur.

1.3 Glacial Lake Outburst Floods (GLOFs)

Glacial lake outburst floods (GLOFs) have caused significant communal, environmental and economic damage to highly populous regions of countries situated around the Himalayan region including Bhutan, China, India, Nepal, and Pakistan. The collective effects of climate change and unplanned land use has increased the frequency of GLOF events since the onset of second half of the twentieth century. Glacier retreat and thinning have occurred because of global warming since the start of the Little Ice Age (Denton and Hughes 1981). Most of the mountain glaciers across the globe have experienced snout retreats (Haeberli et al. 1999) including the Himalayan glaciers (Ageta et al. 2001). However, the recession rates vary across the Himalayas (Murtaza and Romshoo 2017) which is due to the variability in the distribution of glaciers, snowfields, regional climate, and topography across the Himalayas.

Large number of glacial lakes (estimated to be around 5000) have been created in the Himalayan region primarily due to retreat of glaciers (Wang et al. 2015). These glacial lakes vary in size ranging from small meltwater ponds to even huge valley lakes. These Glacier lakes are dammed with unconsolidated weak natural moraines (Budhathoki et al. 2010). When these weak dams break, glacier lake outburst floods (GLOFs) occur and can cause catastrophic impacts. These weak dams pose a potential risk of failure which results in the catastrophic discharge of water under pressure from a glacial lake causing a huge flash flood in the downstream areas commonly known as glacier lake outburst flood (GLOF) (Din et al. 2014). Downstream flash flooding and movement of debris can cause a significant threat to life, property, infrastructure, crops, etc. Many GLOF events have been reported and or studied which have affected human lives and infrastructure (Veh et al. 2020). The studies revealed that the driving factors for such events are weak unconsolidated moraine dams, heavy precipitation, rock failure, and sometimes seismic activities (Vuichard and Zimmermann 1987a, b; Budhathoki et al. 2010; Din et al. 2014) (Fig. 7).

In the Himalayas, the GLOFs have had the highest death toll worldwide (Carrivick and Tweed 2016). Many studies have been carried out on the glacial lakes and the associated GLOF hazards in the Himalayan region. Most of the studies focus on mapping the distribution and dynamics of these moraine-dammed lakes (Maharjan et al. 2018; Schwanghart et al. 2016). However, the complex topography and climate of the Himalayas hamper the field-based work which creates a huge gap in our understanding and mitigating these catastrophic hazards. So, remotely collected data which includes satellite and or Unmanned Aerial Vehicles (UAV), meteorological data, and detailed topographic data help to quantify and estimate parameters associated with these outbursts. These parameters include the ice geometry, moraine dams, possibility of avalanches or landslides, and the water volumes released by these outbursts (Emmer and Vilímek 2013; Fujita et al. 2009; Wang et al. 2012). A glacial lake inventory for identifying potential dangerous glacial lakes in the Himalayas was compiled by International Center for Integrated Mountain Development (ICIMOD) using satellite images, topographic maps, and in-situ data (Mool et al. 2001). Further

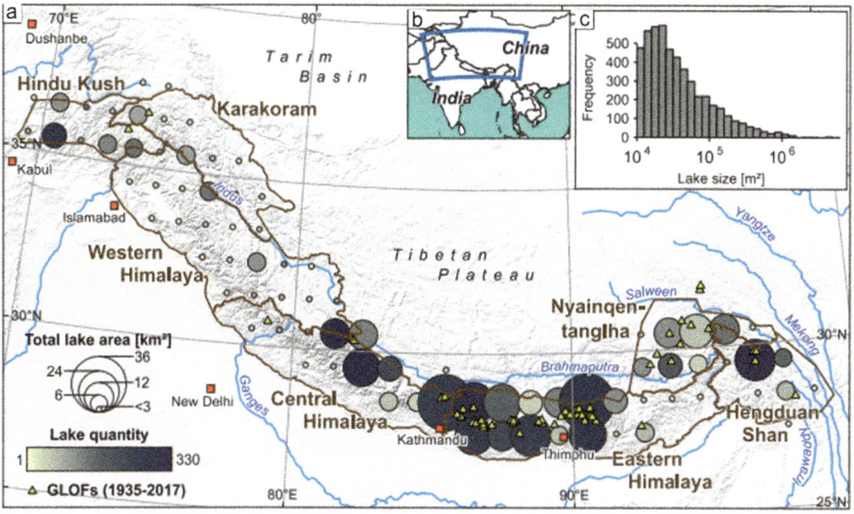

Fig. 7 Moraine-dammed glacier lakes in the Hindu Kush Himalayas represented by bubbles. Yellow triangles represent reported GLOFs. Histogram represents glacier lake areas (Veh et al. 2020)

studies describing the possible trigger mechanisms for GLOF events including failure of moraine dam, mass movement entering a lake in the form of avalanches or landslides, and the wave-erosion of a terminal moraine must be undertaken to make our understanding of GLOFs much better.

Glacier lake outburst floods evidently represent changing climate and mountain cryosphere. The Himalayan region has suffered the highest due to the GLOF hazard and associated catastrophic discharges ow water. A GLOF can have a mean discharge of $\sim$15,600 $m^3\ s^{-1}$, equivalent to monsoonal discharges hundreds of kilometers downstream. Spatially growing glacier lakes and the frequency of lake outbursts primarily determine GLOF hazard. The Eastern Himalayas have GLOF occurrence 3 times than other Himalayan regions followed by Western Himalayas, Hindu Kush, and Karakoram.

Several stakeholders in the Himalayas have identified the need for a regional approach to GLOF risk reduction. Though it is very difficult to control any glacial lake outburst but mitigation strategies can help in decreasing the vulnerability of these elements. Engineering measures such as reinforcing the moraine dam, reducing lake capacity, and ensuring continuous discharge from the glacier lake are the most efficient solutions to prevent a catastrophe. Other measures such as early warning systems, disaster/ emergency management, medical conditions, and disaster insurance can help. Future projections estimate higher rates of ice loss which could generate more unstable moraine-dammed lakes, thus increasing GLOF hazard risks in the region. The increasing population trends, changes in land use land cover patterns, infrastructure development in Himalayan region can highly magnify the damage caused by a GLOF event.

2 Conclusion

Natural disasters like landslides, flash floods, etc., happen very frequently in mountainous regions, especially in Himalayas. They cause huge damage to the property and economy of the region. The impact of natural disasters is an increasing problem in Himalayan region, as illustrated by the recent catastrophic events. Recurring disasters of various magnitudes adversely affect people and ecosystem in Himalayan region. This chapter reviews about various hazards prevalent in mountains terrain. The review highlights that the biggest challenge in the region is organized collection of data along with its management. In order to develop suitable disaster risk reduction policy and early warning systems, there is a need for improved data collection methods, in addition to appropriate procedures for hazard vulnerability analysis. Most of the countries in Himalayan region have high population density with a large proportion of population dependent on agriculture and allied sectors. The reasons behind the intensified vulnerability of Himalayan region to various disasters like landslides, floods, GLOFS, etc., is attributed to existing physical and environmental factors and low economic progress as compared to other parts of the World. In addition, the chapter highlights the integrated use of geospatial techniques along with advanced methods for forecasting, mapping, monitoring and modeling of various hazards in Himalayan region. Over the years several studies have focused on incorporation of machine learning approach in landslide hazard susceptibility mapping. At regional scale, hydrometeorological studies in most of the Himalayan region are not recurrently available for a longer time duration. The amalgamated method for the collection of hydrological data at basin level is important to get the perceptive of runoff generation, predominantly when combined with meteorological data. A detailed inventory of all the extreme events including landslides, flash floods, GLOFs, etc., with attributes like location, elevation, etc., needs to be compiled. Rapid anthropogenic developments have accelerated the impacts of the disasters across the globe, especially in Himalayas. Therefore, there is an urgent need for risk assessment to assist the management of risk, scientifically and rationally, to safeguard against loss of property and life. Risk management must put attention on high-risk zones. It is required that government administrators need to categorize degree of severity and necessity of disaster damage to put forward prevention and control plans in a phased manner. Particularly, for infrastructural and engineering planning purpose in mountainous areas, it should be mandatory for government agencies to carry out comprehensive disaster risk valuation before the construction. Over the years, it has been reported that the human fatalities are on rise due to occurrence of frequent disasters, which arises the need for mitigation measures to be on the priority list in all susceptible areas. The structural measures established by developed countries are impossible to apply in developing countries due to high capital investment. Therefore, in these regions non-structural procedures like early warning systems are more likable, based on a mathematical model that relates the input components to the corresponding output component.

References

Ageta Y, Naito N, Nakawo M, Fujita K, Shankar K (2001) Study project on the recent rapid shrinkage of summer-accumulation type glaciers in the Himalayas, 1997–1999. Bull Glaciol Res 18:45–49

Alfieri L, Burek P, Dutra E, Krzeminski B, Muraro D, Thielen J, Pappenberger F (2013) GloFAS–global ensemble streamflow forecasting and flood early warning. Hydrol Earth Syst Sci 17(3):1161–1175

Ali K, Bajracharya RM, Koirala HL, Begum F (2016) A review of flood risk assessment

Althuwaynee OF, Pradhan B, Park HJ, Lee JH (2014) A novel ensemble bivariate statistical evidential belief function with knowledge-based analytical hierarchy process and multivariate statistical logistic regression for landslide susceptibility mapping. Catena

Anbalagan R (1992) Landslide hazard evaluation and zonation mapping in mountainous terrain. Eng Geol 32(4):269–277

Apollo M (2017) The population of Himalayan regions—by the numbers: past, present and future. In: Contemporary studies in environment and tourism

Arora M, Kumar R, Singh RD, Malhotra J, Kumar N (2016) Analysis of unusual meteorological conditions that led to recent floods in Bhagirathi basin (Uttarakhand Himalayas). Hydrol Sci J 61(7):1239–1244

Bates PD, De Roo APJ (2000) A simple raster-based model for flood inundation simulation. J Hydrol

Bates PD, Dawson RJ, Hall JW, Horritt MS, Nicholls RJ, Wicks J, Hassan MAAM (2005) Simplified two-dimensional numerical modelling of coastal flooding and example applications. Coast Eng 52(9):793–810

Bertram GCL, Burton I, Kates RW, White GF (1979) The environment as hazard. Geograph J

Bouwer LM (2011) Have disaster losses increased due to anthropogenic climate change? Bull Am Meteorol Soc

Bradford, Scott F, Sanders BF (2002) Finite-volume model for shallow-water flooding of arbitrary topography. J Hydraul Eng

Brakenridge R, Anderson E (2006) MODIS-based flood detection, mapping and measurement: the potential for operational hydrological applications. In: Transboundary floods: reducing risks through flood management

Budhathoki KP, Bajracharya OR, Pokharel BK (2010) Assessment of Imja Glacier Lake outburst flood (GLOF) risk in Dudh Koshi River Basin using remote sensing techniques. J Hydrol Meteorol 7(1):75–91

Carrivick, Jonathan L, Tweed FS (2016) A global assessment of the societal impacts of glacier outburst floods. Global and planetary change

Casulli V, Cheng RT (1992) Semi-implicit finite difference methods for three-dimensional shallow water flow. Int J Numer Methods Fluids

Das I, Stein A, Kerle N, Dadhwal VK (2012) Landslide susceptibility mapping along road corridors in the Indian Himalayas using Bayesian logistic regression models. Geomorphology 179:116–125

Denton GH, Hughes TJ (eds) (1981) The last great ice sheets, vol 1. Wiley

Dhar ON, Shobha N (2003) Hydrometeorological aspects of floods in India. Nat Hazards

Dhar S, Rai AK, Nayak P (2017) Estimation of seismic hazard in Odisha by remote sensing and GIS techniques. Nat Hazards

Dikshit A, Sarkar R, Pradhan B, Segoni S, Alamri AM (2020) Rainfall induced landslide studies in Indian Himalayan region: a critical review. Appl Sci (Switzerland)

Dimri AP, Niyogi D, Barros AP, Ridley J, Mohanty UC, Yasunari T, Sikka DR (2015) Western disturbances: a review. In: Reviews of geophysics

Din K, Tariq S, Mahmood A, Rasul G (2014) Temperature and precipitation: GLOF triggering indicators in Gilgit-Baltistan, Pakistan. Pak J Meteorol 10(20)

Dottori F, Todini E (2011) Developments of a flood inundation model based on the cellular automata approach: testing different methods to improve model performance. Phys Chem Earth

Dottori F, Todini E (2013) Testing a simple 2D hydraulic model in an urban flood experiment

Dumbser M, Uwe I, Matteo I (2015) An efficient semi-implicit finite volume method for axially symmetric compressible flows in compliant tubes. Appl Numer Math

El Kadi A, Kamal AP, Emmanuel M (2009) Modelling flash flood propagation in urban areas using a two-dimensional numerical model. In: Natural hazards

Emmer A, Vilímek V (2013) Review article: lake and breach hazard assessment for Moraine-dammed lakes: an example from the Cordillera Blanca. Nat Hazards Earth Syst Sci

Ervine DA, MacLeod AB (1999) Modelling a river channel with distant floodbanks. In: Proceedings of the institution of civil engineers: water, maritime and energy

Ferreira S, Hamilton K, Vincent JR (2011) Nature, socioeconomics and adaptation to natural disasters: new evidence from floods

Fujita K, Sakai A, Nuimura T, Yamaguchi S, Sharma RR (2009) Recent changes in Imja Glacial Lake and its damming moraine in the Nepal Himalaya revealed by in situ surveys and multi-temporal ASTER imagery. Environ Res Lett 4(4):045205

George DL (2011) Adaptive Finite volume methods with well-balanced Riemann solvers for modeling floods in rugged terrain: application to the malpasset dam-break flood (France, 1959). Int J Numer Methods Fluids

Ghimire B, Chen AS, Guidolin M, Keedwell EC, Djordjević S, Savić DA (2013) Formulation of a fast 2D urban pluvial flood model using a cellular automata approach. J Hydroinf 15(3):676–686

Ghimire M (2011) Landslide occurrence and its relation with terrain factors in the Siwalik Hills, Nepal: case study of susceptibility assessment in three basins. Natural Hazards

Ghosh S, Carranza EJM, van Westen CJ, Jetten VG, Bhattacharya DN (2011) Selecting and weighting spatial predictors for empirical modeling of landslide susceptibility in the Darjeeling Himalayas (India). Geomorphology 131(1–2):35–56

Ghosh T, Bhowmik S, Jaiswal P, Ghosh S, Kumar D (2020) Generating substantially complete landslide inventory using multiple data sources: a case study in Northwest Himalayas, India. J Geol Soc India 95(1):45–58

Glickman TS (1992) Acts of god and acts of man: recent trends in natural disasters and major industrial accidents

Guhathakurta P, Sreejith OP, Menon PA (2011) Impact of climate change on extreme rainfall events and flood risk in India. J Earth Syst Sci

Gupta, Anil K, Nair SS (2011) Urban floods in Bangalore and Chennai: risk management challenges and lessons for sustainable urban ecology. Curr Sci

Haeberli W, Regula F, Martin H, Max M (1999) On rates and acceleration trends of global glacier mass changes. Geografiska Annaler, Ser Phys Geogr

Haider S, André P, Robert M, Jean YC (2003) Urban flood modelling using computational fluid dynamics. In: Proceedings of the institution of civil engineers: water and maritime engineering

Hewitt K (2005) The Karakoram anomaly? Glacier expantion and the 'elevation effect,' Karakoram Himalaya." Mountain Research and Development

Hirabayashi Y, Mahendran R, Koirala S, Konoshima L, Yamazaki D, Watanabe S, Kim H, Kanae S (2013) Global flood risk under climate change. Nat Clim Chang 3(9):816–821

Horritt MS, Mason DC, Cobby DM, Davenport IJ, Bates PD (2003) Waterline mapping in flooded vegetation from airborne SAR imagery. Remote Sens Environ 85(3):271–281

Houze RA, Rasmussen KL, Medina S, Brodzik SR, Romatschke U (2011) Anomalous atmospheric events leading to the summer 2010 floods in Pakistan. Bull Am Meteor Soc 92(3):291–298

Hunter NM, Bates PD, Neelz S, Pender G, VillanuevaI, Wright NG, Liang D, Mason DC (2008) Benchmarking 2D hydraulic models for urban flooding. In: Proceedings of the institution of civil engineers-water management, vol 161(1). Thomas Telford Ltd., pp 13–30

Hunter NM, Bates PD, Horritt MS, Wilson MD (2007) Simple spatially-distributed models for predicting flood inundation: a review. Geomorphology

IRDR (2014) Peril classificationa and hazard Terminologyeijing (IRDR DATA Publication No. 1). Integrated research on disaster risk

Jayawardena AW (2015) Hydro-meteorological disasters: causes, effects and mitigation measures with special reference to early warning with data driven approaches of forecasting. Procedia IUTAM 17(2013):3–12

Jeelani G, Feddema JJ, Van Der Veen CJ, Stearns L (2012) Role of snow and glacier melt in controlling river hydrology in Liddar watershed (Western Himalaya) under current and future climate. Water Resour Res 48(12):1–16

Kabenge M, Elaru J, Wang H, Li F (2017) Characterizing flood hazard risk in data-scarce areas, using a remote sensing and GIS-based flood hazard index. Nat Hazards

Kaku K, Alexander H (2013) Sentinel Asia: a space-based disaster management support system in the Asia-Pacific Region. Int J Disaster Risk Reduction

Kanungo DP, Arora MK, Gupta RP, Sarkar S (2008) Landslide risk assessment using concepts of danger pixels and fuzzy set theory in Darjeeling Himalayas. Landslides 5(4):407–416

Kappes MS, Keiler M, von Elverfeldt K, Glade T (2012) Challenges of analyzing multi-hazard risk: a review. In: Natural Hazards

Kazakis N, Kougias I, Patsialis T (2015) Assessment of flood hazard areas at a regional scale using an index-based approach and analytical hierarchy process: application in Rhodope-Evros region, Greece. Sci Total Environ

Komolafe AA, Awe BS, Olorunfemi IE, Oguntunde PG (2020) Modelling flood-prone area and vulnerability using integration of multi-criteria analysis and HAND model in the Ogun River Basin. Nigeria. Hydrol Sci J 65(10):1766–1783

Komolafe AA, Srikantha H, Ram A (2018) Methodology to assess potential flood damages in urban areas under the influence of climate change. Nat Hazards Rev

Komolafe AA, Srikantha H, Ram A, Jean FV (2019) Comparative analyses of flood damage models in three Asian countries: towards a regional flood risk modelling. Environ Syst Decis

Kumar R, Anbalagan R (2015) Landslide susceptibility zonation in part of Tehri reservoir region using frequency ratio, fuzzy logic and GIS. J Earth Syst Sci 124(2):431–448

Kumar V, Gupta V, Jamir I (2018) Hazard evaluation of progressive Pawari landslide zone, Satluj valley, Himachal Pradesh. India. Nat Hazards 93(2):1029–1047

Kumar V, Gupta V, Sundriyal YP (2019) Spatial interrelationship of landslides, litho-tectonics, and climate regime, Satluj valley. Northwest Himalaya. Geol J 54(1):537–551

Kwak YJ (2017) Nationwide flood monitoring for disaster risk reduction using multiple satellite data. ISPRS Int J Geo-Inf

Lal P et al (2020) Evaluating the 2018 extreme flood hazard events in Kerala, India. Remote Sens Lett 11(5):436–445

Lehner B, Döll P, Alcamo J, Henrichs T, Kaspar F (2006) Estimating the impact of global change on flood and drought risks in Europe: a continental, integrated analysis. Clim Change

LeVeque, Randall J (1997) Wave propagation algorithms for multidimensional hyperbolic systems. J Comput Phys

Liu L, Liu Y, Wang X, Yu D, Liu K, Huang H, Hu G (2015) Developing an effective 2-D urban flood inundation model for city emergency management based on cellular automata. Nat Hazard 15(3):381–391

Ma X, Jianchu X, Meine van N (2010) Sensitivity of streamflow from a himalayan catchment to plausible changes in land cover and climate. Hydrol Processes

Madhura RK, Krishnan R, Revadekar JV, Mujumdar M, Goswami BN (2014) Changes in western disturbances over the Western Himalayas in a warming environment. Clim Dyn

Maharjan SB et al (2018) The status of glacial lakes in the Hindu Kush Himalaya 70

Mandal S, Mandal K (2018) Modeling and mapping landslide susceptibility zones using GIS based multivariate binary logistic regression (LR) model in the Rorachu river basin of eastern Sikkim Himalaya. India. Modeling Earth Syst Environ 4(1):69–88

Martha TR, Kumar KV (2013) September, 2012 landslide events in Okhimath, India—an assessment of landslide consequences using very high resolution satellite data. Landslides 10(4):469–479

Martha TR, Kamala P, Jose J, Kumar KV, Sankar GJ (2016) Identification of new landslides from high resolution satellite data covering a large area using object-based change detection methods. J Indian Soc Remote Sens 44(4):515–524

Martha TR, Kerle N, Jetten V, van Westen CJ, Kumar KV (2010) Characterising spectral, spatial and morphometric properties of landslides for semi-automatic detection using object-oriented methods. Geomorphology 116(1–2):24–36

Martha TR, Kerle N, Van Westen CJ, Jetten V, Kumar KV (2012) Object-oriented analysis of multi-temporal panchromatic images for creation of historical landslide inventories. ISPRS J Photogramm Remote Sens 67:105–119

Martha TR, Roy P, Govindharaj KB, Kumar KV, Diwakar PG, Dadhwal VK (2015) Landslides triggered by the June 2013 extreme rainfall event in parts of Uttarakhand state. India. Landslides 12(1):135–146

Martha TR, van Westen CJ, Kerle N, Jetten V, Kumar KV (2013) Landslide hazard and risk assessment using semi-automatically created landslide inventories. Geomorphology 184:139–150

Mason DC et al (2012) Near real-time flood detection in urban and rural areas using high-resolution synthetic aperture radar images. IEEE Trans Geosci Remote Sens

Mathew J, Jha VK, Rawat GS (2007) Weights of evidence modelling for landslide hazard zonation mapping in part of Bhagirathi valley, Uttarakhand. Curr Sci, 628–638

Mignot E, Paquier A, Haider S (2006) Modeling floods in a dense urban area using 2D shallow water equations. J Hydrol

Milly PCD, Wetherald RT, Dunne KA, Delworth TL (2002) Increasing risk of great floods in a changing climate. Nature

Mohapatra PK, Singh RD (2003) Flood management in India. Nat Hazards

Mool P, Bajracharya SR, Joshi SP (2001) Inventory of glaciers, glacial lakes, and glacial lake outburst floods: Monitoring and early warning systems in the Hindu Kush-Himalayan region—Nepal. ICIMOD, Kathmandu

Mukherjee F, Deepika S (2020) Detecting flood prone areas in Harris County: a GIS based analysis. GeoJournal

Murtaza KO, Romshoo SA (2017) Recent Glacier Changes in the Kashmir Alpine Himalayas, India. Geocarto Int 32(2):188–205

Néelz S, Pender G, Britain G (2009) Environment agency desktop review of 2D hydraulic modelling packages

Nepal S, Shrestha AB (2015) Impact of climate change on the hydrological regime of the Indus, Ganges and Brahmaputra river basins: a review of the literature. Int J Water Resourc Dev

Nkwunonwo UC, Whitworth M, Baily B (2020) A review of the current status of flood modelling for urban flood risk management in the developing countries. Sci Afr 7: e00269

Ordoyne C, Friedl MA (2008) Using MODIS data to characterize seasonal inundation patterns in the Florida everglades. Remote Sens Environ

Pandey VK, Sharma MC (2017) Probabilistic landslide susceptibility mapping along Tipri to Ghuttu highway corridor, Garhwal Himalaya (India). Remote Sens Appl Soc Environ 8:1–11

Patro S, Chatterjee C, Mohanty S, Singh R, Raghuwanshi NS (2009) Flood inundation modeling using MIKE FLOOD and remote sensing data. J Indian Soc Remote Sens 37(1):107–118

Peethambaran B, Anbalagan R, Shihabudheen KV, Goswami A (2019) Robustness evaluation of fuzzy expert system and extreme learning machine for geographic information system-based landslide susceptibility zonation: a case study from Indian Himalaya. Environ Earth Sci 78(6):1–20

Pellicani R, Argentiero I, Spilotro G (2017) GIS-based predictive models for regional-scale landslide susceptibility assessment and risk mapping along road corridors. Geomatics Nat Hazards Risk

Pham BT, Bui DT, Prakash I, Dholakia MB (2017) Hybrid integration of multilayer perceptron neural networks and machine learning ensembles for landslide susceptibility assessment at Himalayan area (India) using GIS. CATENA 149:52–63

Pham BT, Nguyen-Thoi T, Qi C, Van Phong T, Dou J, Ho LS, Van Le H, Prakash I (2020) Coupling RBF neural network with ensemble learning techniques for landslide susceptibility mapping. Catena 195:104805

Pham BT, Prakash I, Singh SK, Shirzadi A, Shahabi H, Bui DT (2019) Landslide susceptibility modeling using reduced error pruning trees and different ensemble techniques: hybrid machine learning approaches. CATENA 175:203–218

Pourghasemi HR, Yansari ZT, Panagos P, Pradhan B (2018) Analysis and evaluation of landslide susceptibility: a review on articles published during 2005–2016 (periods of 2005–2012 and 2013–2016). Arab J Geosci 11(9):1–12

Ramakrishnan D, Singh TN, Verma AK, Gulati A, Tiwari KC (2013) Soft computing and GIS for landslide susceptibility assessment in Tawaghat area, Kumaon Himalaya. India Nat Hazards 65(1):315–330

Reichenbach P, Rossi M, Malamud BD, Mihir M, Guzzetti F (2018) A review of statistically-based landslide susceptibility models. Earth Sci Rev 180:60–91

Rinaldi, Pablo R, Dalponte DD, Vénere MJ, Clausse A (2007) Cellular automata algorithm for simulation of surface flows in large plains. Simul Modelling Pract Theory

Rozalis S, Efrat M, Yoav Y, Colin P (2010) Flash flood prediction using an uncalibrated hydrological model and radar rainfall data in a mediterranean watershed under changing hydrological conditions. J Hydrol

Sanyal J, Lu XX (2005) Remote sensing and GIS-based flood vulnerability assessment of human settlements: a case study of gangetic West Bengal, India. Hydrol Processes

Sarkar S, Roy AK, Raha P (2016) Deterministic approach for susceptibility assessment of shallow debris slide in the Darjeeling Himalayas, India. CATENA 142:36–46

Schwanghart W, Worni R, Huggel C, Stoffel M, Korup O (2016) Uncertainty in the Himalayan energy–water nexus: estimating regional exposure to glacial lake outburst floods. Environ Res Lett 11(7):074005

Sharma S, Mahajan AK (2019) A comparative assessment of information value, frequency ratio and analytical hierarchy process models for landslide susceptibility mapping of a Himalayan watershed, India. Bull Eng Geol Env 78(4):2431–2448

Shi P (2019) Disaster risk science. Springer

Shrestha M (2008) Impacts of floods in South Asia. J South Asia Disaster Stud

Singh N, Ranade A (2010) Determination of onset and withdrawal dates of summer monsoon across India using NCEP/NCAR re-analysis

Singh P, Lars B (2004) Hydrological sensitivity of a large himalayan basin to climate change. Hydrol Processes

Skilodimou HD, Bathrellos GD, Chousianitis K, Youssef AM, Pradhan B (2019) Multi-hazard assessment modeling via multi-criteria analysis and GIS: a case study. Environ Earth Sci 78(2):47

Soares-Frazão S, Julien L, Vincent G, Yves Z (2008) Two-dimensional shallow-water model with porosity for urban flood modelling. J Hydraul Res

Szewrański S, Chruściński J, Kazak J, Świąder M, Tokarczyk-Dorociak K, Żmuda R (2018) Pluvial flood risk assessment tool (PFRA) for rainwater management and adaptation to climate change in newly urbanised areas. Water 10(4):386

Theilen-Willige B (2010) Detection of local site conditions influencing earthquake shaking and secondary effects in Southwest-Haiti using remote sensing and GIS-methods. Nat Hazards Earth Syst Sci

Thilagavathi G, Tamilenthi S, Ramu C, Baskaran R (2011) Application of Gis in flood hazard zonation studies in Papanasam Taluk, Thanjavur District, Tamilnadu. Adv Appl Sci Res

Turner AK, Schuster RL (1996) Landslides: investigation and mitigation. Special Report 247. Trans. Res. Board, National Academy Press, Washington, DC

Twele A, Wenxi C, Simon P, Sandro M (2016) Sentinel-1-based flood mapping: a fully automated processing chain. Int J Remote Sens

UNISDR (2009) Terminology on disaster risk reduction. United Nations Office for Disaster Risk Reduction, Report

Vamsee AM, Kamala P, Martha TR, Kumar KV, Amminedu E (2018) A tool assessing optimal multi-scale image segmentation. J Indian Soc Remote Sens 46(1):31–41

Veh G, Oliver K, Ariane W (2020) Hazard from himalayan glacier lake outburst floods 117(2): 907–12

Vuichard D, Zimmermann M (1987) The 1985 catastrophic drainage of a Moraine-Dammed Lake, Khumbu Himal, Nepal: cause and consequences. Mt Res Dev 7(2):91–110

Vuichard D, Zimmermann M (1987) The 1985 catastrophic drainage of a moraine-dammed lake, Khumbu Himal, Nepal: cause and consequences. Mt Res Dev, 91–110

Wang W, Xiang Y, Gao Y, Lu A, Yao T (2015) Rapid expansion of glacial lakes caused by climate and glacier retreat in the Central Himalayas. Hydrol Process 29(6):859–874

Wang X, Liu S, Ding Y, Guo W, Jiang Z, Lin J, Han Y (2012) An approach for estimating the breach probabilities of moraine-dammed lakes in the Chinese Himalayas using remote-sensing data. Nat Hazard 12(10):3109–3122

Ward PJ, Jongman B, Salamon P, Simpson A, Bates P, Groeve De T, Muis S, De Perez EC, Rudari R, Trigg MA, Winsemius HC (2015) Usefulness and limitations of global flood risk models. Nat Clim Chang 5(8):712–715

Webster PJ, Holland GJ, Curry JA, Chang HR (2005) Atmospheric science: changes in tropical cyclone number, duration, and intensity in a warming environment. Science

Wilby, Robert L, Rod K (2012) Adapting to flood risk under climate change. Progr Phys Geogr

Winsemius HC, Aerts JC, Beek Van LP, Bierkens MF, Bouwman A, Jongman B, Kwadijk JC, Ligtvoet W, Lucas PL, Van Vurren DP, Ward PJ (2016) Global drivers of future river flood risk. Nat Clim Chang 6(4):381–385

Wisner B, Blaikie P, Cannon T, Davis I (2014) At risk: natural hazards, people's vulnerability and disasters. In: At risk: natural hazards peoples vulnerability and disasters

Wu H, Adler RF, Hong Y, Tian Y, Policelli F (2012) Evaluation of global flood detection using satellite-based rainfall and a hydrologic model. J Hydrometeorol 13(4):1268–1284

Wu H, Adler RF, Tian Y, Huffman GJ, Li H, Wang J (2014) Real-time global flood estimation using satellite-based precipitation and a coupled land surface and routing model. Water Resour Res 50(3):2693–2717

Wu H, Huang M, Tang Q, Kirschbaum DB, Ward P (2016) Hydrometeorological hazards: monitoring, forecasting, risk assessment, and socioeconomic responses

Yu D, Lane SN (2006) Urban fluvial flood modelling using a two-dimensional diffusion-wave treatment, Part 1: mesh resolution effects. Hydrological Processes

Landslides in the Himalayas—Causes and Consequences

Himalayan Landslides–Causes and Evolution

Sandeep Singh, Anand Joshi, Anamika Sahu, R. Arun Prasath, Saurabh Sharma, and Chandra Shekhar Dwivedi

Abstract The Himalayas has been characterized by many superlatives, viz. youngest mountain chain, the highest peak, home of severe earthquakes, and the highest cases of landslides. The landslides are inevitable due to the presence of fragile rocks, the presence of major tectonic boundaries, and the activities along with them due to the northward movement of the Indian Plate; Earthquakes of high magnitude. Both geological and historical records indicate landslides devastating nature, causing a large scale of destruction and losses. There has been an emphasis on monitoring landslides with various modern techniques for systematic studies to highlight the landslide's extent and effect in suggesting proper remedial measures.

Keywords Himalayas · Landslides · Earthquakes · Monitoring of landslides

1 Introduction

Landslides are considered one of the most significant natural hazards worldwide. The total number of fatal landslides in the world indicates that nearly 75% of total occur in Asian countries, out of which substantial numbers have occurred in Himalayas (Froude and Petley 2018). The Himalayas regions belong to moderate to very high global hotspot landslide hazard zonation with a high Mortality rate for expected annual mortality risk of landslides worldwide (Nadim et al. 2006; Yang et al. 2015). The Himalayas represent a rugged topography zone (Fig. 1), high-intensity rainfall, and the rain shadow zone, along with the high magnitude of the earthquake (Fig. 2).

S. Singh (✉) · A. Joshi · A. Sahu · S. Sharma
Department of Earth Sciences, Indian Institute of Technology Roorkee, Roorkee 247667, India
e-mail: sandeep.singh@es.iir.ac.in

R. Arun Prasath
Seismology and Geoscience Division, Ministry of Earth Sciences, Prithvi Bhavan, Lodhi Road, New Delhi 110003, India

C. S. Dwivedi
Department of Geoinformatics, Central University of Jharkhand, Ranchi 835205, India

S. Kanga et al. (eds.), *Disaster Management in the Complex Himalayan Terrains*, Geography of the Physical Environment,
https://doi.org/10.1007/978-3-030-89308-8_3

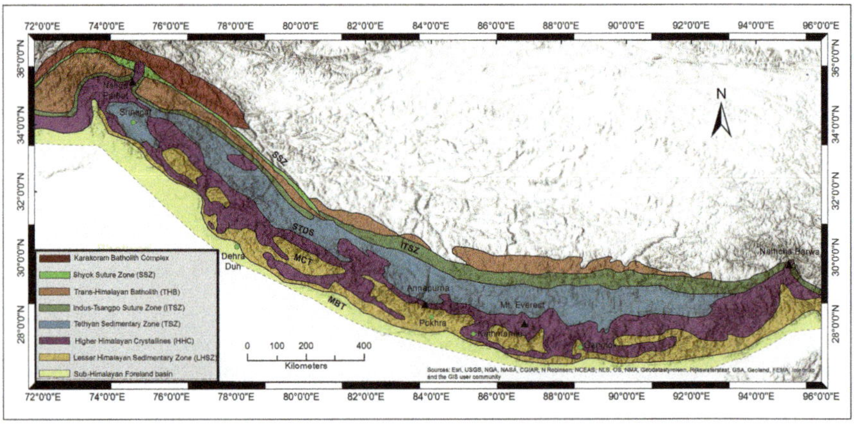

Fig. 1 Simplified regional geological framework of the Himalayas in Plate tectonic framework. Higher Himalayan Crystallines (HHC) have been redefined by few workers as Great Himalayan Sequence (GHS) Abbreviations: SSZ-Shyok Suture Zone. ITSZ-Indus Tsangpo Suture Zone. THSZ/STDS-Trans Himadri Shear Zone/South Tibetan Detachment System (STDS). MCT-Main Central Thrust. MBT-Main Boundary Thrust. After Singh (2019)

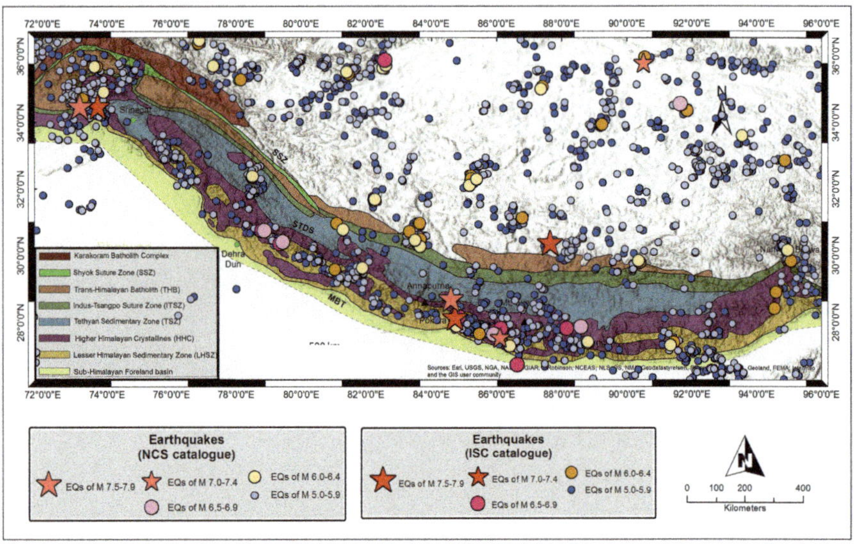

Fig. 2 Location of earthquakes and magnitude based on ISC-catalog (ISC 2020; last accessed: 29th Dec 2020) and NCS-catalog (NCS 2020; last accessed: 29th Dec 2020) in the Himalayas. The base map is after Singh (2019)

Landslides, debris flow, soil erosions, and other mass wasting processes characterize the Himalayas. The erosions have led to the formation of Himalayan Foreland basins, which have been described by the Sub- and Lesser Himalayan zone and the Indo-Gangetic Plain (Singh and Singh 2020).

The reason for substantial numbers of landslides in the Himalayas is the presence of vulnerable rocks, steep slopes, and building up of strain due to the northward movement of the Indian Plate beneath the Asian Plate (Singh 2019 and references therein). The continuous northward movement of the Indian Plate and its collision with the Eurasian Plate is causing earthquakes and other neo-tectonic activities in this part of the Indian subcontinent. Weak, incompetent rocks are more likely to cause landslides than strong, competent rocks; similarly, the general steeper slope has a greater chance of land sliding than the gentle slope. The frequency of landslide increases many folds once the equilibrium is disturbed in the highly undulating mountainous terrain.

During the monsoon (summer and winter), landslides and related natural disasters affect human life in the Himalayan region. Most of the human settlements are situated in the paleo-landslide zone due to the presence of fertile soils. Rainfall raises pore-water pressure in slope materials, which can trigger landslides. Earthquakes, increased overburden, change in slope gradient, mining, hill cutting, constructions, human activities, etc., can also trigger landslides.

Dikshit et al. (2020) analyzed the landslide studies based on the web of science across the states in the Indian Himalayan region. They worked out that the studies are quite biased toward the Uttrakhand region consisting of 51%, followed by West Bengal (15%), Himachal Pradesh (13%), Sikkim (10%), Jammu and Kashmir (5%), Arunachal Pradesh (3%), Mizoram (2%), Manipur (1%). It also indicates that northeastern Himalaya has very few to almost none studies.

2 Type of Landslides

Landslides often occur on hillslopes, which pull the soil and rock downslope; this only occurs when the developed stress in the rock mass exceeds the strength of the hill slope material (Hyndman and Hyndman 2009). It shows significant variability in terms of its typology along with kinematics and geometric variabilities. The landslide classification is mainly based on the type of material involved along with the type of movement along any slope (Cruden and Varnes 1996). There could be more than one type of movement and the cause of triggering the landslide. Varnes (1978) made the first attempt to classify the landslide and has been widely used and even adopted by Landslide Committee, Highway Research Board, Washington (Thakur 1996). They recognized the movement of soil, rock, debris, or earth downslope either by fall, topple, slide, spread, or flow.

3 Himalayas and Lithotectonic Units

The Himalayas results from the continent–continent collision between the Indian Plate and the Asian Plate no later than 57 Ma (Leech et al. 2005, 2007). It forms a fascinating, spectacular, modified sculptured landscape between two syntaxial bends knows as Namche Barua (7782 m) in the east and Nanga Parbat (8125 m) in the west, where the range takes a sharp turn toward the south (Figs. 1 and 2—Singh 2019 and references therein). The collision resulted in the Himalayas formation with extreme and intense crustal shortening and upliftment of the world's highest and youngest mountain chain.

From north to south, the Himalayas can be divided into five lithotectonic units with distinct characters (also see Jain et al. 2002; Yin 2006) exposed along the E-W strike of the Himalayan orogeny, they are:

1. The Tethyan Sedimentary Zone (TSZ)
2. The Higher Himalayan Crystallines (HHC)
3. The Lesser Himalayan Sedimentary Zone (LHSZ)
4. The Sub-Himalayan Foreland Basin (SHFB)
5. The Indo-Gangetic Plain (IGP).

The **Tethyan Sedimentary Zone (TSZ)** is the northernmost part of the Indian Plate lithotectonic units of the Himalayas. It consists of mildly deformed to almost undeformed sedimentary sequence very prone to landslides, but due to rain shadow zone and less habitation, there is less threat for landslides. The South Tibetan Detachment System (STDS) separates these rocks in the south from Higher Himalayan Crystallines (HHC). The rocks are composed of shale, limestone, and sandstone, ranging from Neoproterozoic to Eocene.

The **Higher Himalayan Crystallines (HHC)** is composed of crystalline rocks and are divided into two thrust sheets; the lower portion between Main Central Thrust (MCT) and the Vaikrita Thrust (VT) called as Munsiari Group of rocks and the upper part between Vaikrita Thrust (VT) and STDS known as Vaikrita Group of rocks. Both these packages form 15 to 20 km thick high-grade metamorphic rocks all along Himalayan orogeny with varying thickness. The rocks are made up of schist, gneisses, quartzite, marble, migmatites, and granite bodies of various ages (see Singh and Jain 2003; Singh 2020).

The **Lesser Himalayan Sedimentary Zone (LHSZ)** is exposed in two zones and can be classified as Inner Lesser Himalayan Zone (ILH), occurring in window structure, and Outer Lesser Himalayan Zone (OLH) bounded between Main Boundary Thrust (MBT) in the south and Main Central thrust (MCT) in the north which separates them from HHC. The rocks are mostly low-grade sedimentary sequences and made up of mostly unfossiliferous shale, sandstone, limestone, dolomite, slate, phyllite, schist, and quartzite.

The **Sub-Himalayan Foreland Basin (SHFB),** also known as **Shiwalik Belt,** is exposed in the south of LHSZ between Main Boundary Thrust (MBT) in the north and Himalayan Frontal Thrust (HFT) in the south. The rocks are consisting of mudstone, siltstone, shale, sandstone, and conglomerate. The rocks of this region are prone to landslide because of the loose nature of rocks and monsoonal activities.

The **Indo-Gangetic Plain (IGP)** is a compressional tectonics product between the Early Miocene and Middle Miocene and attained present-day configuration (Singh 1996). The GAP accumulated eroded sediments from the various Himalayas and Peninsular India lithologies during Cenozoic time (Shukla et al. 2012). GAP geometry is controlled by flexural subsidence related to the foreland basin character of the Himalayas having depo-center close to the front (Mungier and Huyghe 2006). Geophysical data suggest transverse ridges and saddles (e.g., Delhi Haridwar Ridge; Dholpur Saddle; Faizabad Ridge; Meja Saddle). The northern depressions (e.g., the Sharda Depression; the Bahraich Depression; the Gandak Depression—Srinivas et al. 2013; Mangalik et al. 2015 and references therein) having graben-like structures that go as deep as ~4 km (Manglik et al. 2015; Singh and Singh 2020).

4 Causes of Landslides

Rock type and its effects: Each rock has a distinct character in terms of chemistry, mineralogy, and textural attributes. Rocks' primary character is a significant factor in determining the strength of rocks. Simultaneously, secondary discontinuities and the surface layer of weathered material are the main factors for the landslide occurrences in any area. Secondary discontinuities are faults, joints, and bedding planes. Rock failure causing landslides mainly depends on the geometry and mechanical properties of secondary discontinuities with slope geometry.

Tectonic Boundaries and their effects: In Himalayas, major tectonic boundaries like South Tibetan Detachment System (STDS), Main Central Thrust (MCT), Main Boundary Thrust (MBT), Himalayan Frontal Thrust (HFT) and their associated splays are playing a critical role in the stress accommodation along the whole of the orogeny. The movements along these major tectonic boundaries and their splays are responsible for generating earthquakes in the region. The distances from these major tectonic boundaries affect the severity of landslides.

Earthquakes and their effects: Strong ground motion associated with earthquakes weaken slope material causing co-seismic landslides. The landslides contribute about 20–25% of the losses due to earthquakes and are known as earthquake-triggered landslides. The probability that landslides will occur in landslide-prone areas generally increases if an earthquake strikes that region. An earthquake occurs along a plane known as the fault plane. Earthquakes happen when two rock blocks slip past each other along the fault, causing stress to buildup. When the stress exceeds the moving blocks' frictional energy, earthquakes occur, and seismic waves are formed. The spot where the rock breaks in the subsurface and is the nucleation point of the released energy is the hypocentre.

There are many accompanying effects with large earthquakes, viz. landslides, tsunami, conflagrations, etc., (Lowrie 2007). In the mountainous region, the earthquake's associated hazard triggers landslide, which can cause devastation even in the areas far away from the epicenter. Highland and Bobrowsky (2008) defined landslide as the downslope movement of rock, debris, or soil material under the impact of gravity. The downslope movement of rock or soil can be either rotational or translational depending on the surface of rupture. The well-designed structures and buildings are affected because of the local soil response during a landslide. The Indian Himalayas lies in the seismic zone V and IV, making it more vulnerable to larger earthquakes (Indian Standard, I.S. 2002). The ground-shaking due to earthquakes cause fluidization and liquefaction, causing landslides along the slopes. Also, the ground-shaking loosens the rock material, which causes rockfall and rock topple. The minimum earthquake magnitude required to generate a landslide is M_L ~ 4 (Keefer 1984).

There are numerous examples of moderate to great earthquakes in the Himalayas; which are Assam (1897), Kangra (1905), Bihar-Nepal (1934), Shillong (1950), Bihar-Nepal (1988), Uttarkashi (1991), Chamoli (1999), Kashmir (2005), Central Nepal (2015), etc. These great earthquakes are responsible for the disaster and casualties due to the slope failures initiated by these earthquakes (Prakash 2013; Prasath et al. 2019 and references therein). The landslides that occurred due to these earthquakes have been studied to understand the conditions like tectonic, morphological, lithological, etc., which play a vital role in the landslides' occurrences. The possible earthquakes of high peak ground acceleration 0.4 and 0.25 g with a return period of 500 years increase the landslide activities on hillslopes.

Slope stability is the resistance offered by any rock slope or inclined soil to the failure by sliding or collapsing. The slope stability analysis is mainly based on traditional methods, which can be grouped as kinematic analysis, limit equilibrium analysis, and rockfall simulators (Eberhardt 2003). Limit equilibrium methods examine the equilibrium of a landmass tending to slide on the slope under gravity influence. The assumption in these methods is that the shear strengths of the material along the surface under failure is regulated by linear or non-linear relationships between shear strength and the normal stress on the surface under failure (EM 1110-2-1902). The Mohr–Coulomb criterion depicted in Fig. 3 presumes that a yield occurs when the shear stress acting on a surface reaches up to a linearly dependent on the normal stress in the same plane. The straight line touching these Mohr's circles is known as the yield line (Hibbitt 2004).

Paleo-landslides and their effects: Pre-historic landslides are very common in the Himalayas. The human settlements are concentrated within the paleo-landslide zone because of the presence of soil and fertile land. They are the vulnerable center of landslides and can be identified by landform and other geomorphic features.

In the Himalayas, paleo-landslide are often associated with the nick point that is very prominent on the river's longitudinal profile and indicates rejuvenation in the past. A narrow valley characterizes these nick points with a broad valley in the upstream direction and the development of lacustrine deposits in the broad valley due to the filling of paleo-lakes materials (Singh and Jain 2007). The lacustrine deposits also record sesimites (paleoseismic deformational structures) indicating landslides' development due to seismic activities in the nearby area (Singh and Jain 2007 and references therein).

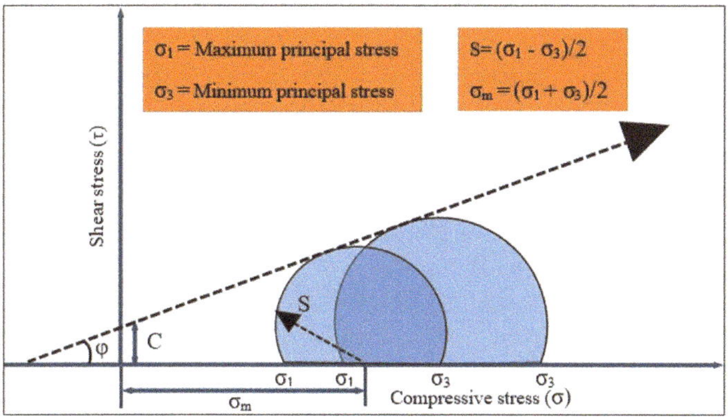

Fig. 3 Linear Mohr–Coulomb failure criterion (Su et al. 2016)

Anthropogenic effects: The anthropogenic intervention aggravated the situation of pre-existing conditions of landslide occurrences. It includes modification/alteration of (i) topography; (ii) water circulation both underground and surface; (iii) land use. These effects generally incorporate deforestation, urbanization, slope support removal during road cutting, mining with blasting, and heavy traffic movement.

5 Monitoring of Landslides

The best possible result from monitoring can be achieved only by a proper understanding of the landslide in geology, geophysics, geomorphology, geohydrology, surveying (aerial extent), and civil engineering aspects. Different methodologies for assessing landslide are influenced by the scale of analysis, input data availability, and required details. For landslide study, it is essential to prepare several thematic maps. The choice includes geological informations in the forms of lithology, structures (foliations, fold, fault, joints, shear zone, etc.), slope angles, relative relief, hydrological details, land use, and land cover. Once they are prepared by overlaying these thematic maps, vulnerable areas can be identified. For generating these maps, various techniques are being used to achieve near realistic pictures, which is useful for future planning.

Applications of geodetic techniques, information, and geospatial technology such as remote sensing and geographic information system (GIS) help monitor the landslides. They include (a) landslide susceptible analysis (Probabilistic approach and/or Deterministic approach)—which include (i) Quantitative method (direct mapping and indirect mapping), (ii) Quantitative method/Data-driven method (Bivariate statistics, Multivariate statistics, Artificial neural network); (b) Runout modeling—it includes volume-based model and dynamic model; (c) landslide monitoring and early warning.

Photogrammetric techniques: Photogrammetry techniques involves the interpretation of aerial photographs (orthophotos) of multi-years. It reveals quantitative information on surface characteristics and indicates geomorphological changes and position size and shape of the landslides (Linder 2009).

Remote Sensing or satellite techniques with space-derived information: (a) Synthetic Aperture Radar (SAR) images, (b) Interferometric Synthetic Aperture Radar (InSAR). An interferometric difference of 2 phase images of the same area at a different time to detect ground motion. Combining many interferograms produce a mean velocity map in which the color shows the ground's speed and indicates sudden movement, if there is any.

Ground-Based Conventional Surveying techniques: (a) Triangular leveling (b) Monitoring by total station. Different surveying techniques are instrumental in the monitoring of landslides, which can include single point positioning (SPP), precise point positioning (PPP) with pseudo kinematic, and real-time kinetic strategies (Tiwari et al. 2018 and references therein).

GPS techniques: The Global Positioning System (GPS) determines the precise determination of point coordinates. Compared to classical surveying, it allows faster and similar accuracy of data acquisition for landslide studies. It can have either the Fast Static (FS) method or Real-Time Kinematic (RTK) techniques (Gilli et al. 2000).

Geotechnical techniques: Sensor-based monitoring (a) Extensiometer: (measures the axial displacement between several reference points in the same measurement axis. They can be installed in boreholes or surfaces. There is also a wire extensometer, typically measuring baseline up to 80 m in length with an accuracy of $\pm$0.3 mm per 30 m. The accuracy gets also affected by temperature. (b) Inclinometer: These are instruments installed in boreholes drilled within the landslide zones. They measure the curvature of the initially straight borehole casing. (c) Piezometer: It measures the pore-water pressure of the landslide zone. Threshold values can be defined to warn. (d) strain meter (e) pressure cell (f) Tiltmeter (g) crack meter (h) geophones: it measures the vibration associated with movement. They can detect landslides based on frequency composition, amplitude, and duration of the vibration signal.

6 Conclusions

- Apart from catastrophic landslides, many small-scale slope failures cause the loss of productive land, which goes unnoticed.
- There is a various cause of landslide occurrence in the Himalayas. It is important to analyze different triggering mechanisms to characterize individual landslides.
- A rapid rise of anthropogenic activities also contributes to the triggering of landslides in the Himalayas.
- It is, therefore, require to undertake systematic studies of landslides, which is still in their infancy.

- It is necessary to prepare zoning maps of landslide-prone areas through geological and geo-technical studies.
- Monitoring of landslides' activities with modern techniques is vital for future planning along with remedial measures.
- The remedial measures involve reforestation, proper drainage, erecting protection structures, and reducing slope angles.

References

Cruden DM, Varnes DJ (1996) Landslides: investigation and mitigation. Chapter 3-Landslide types and processes. Transportation research board special report, 247

Dikshit A, Sarkar R, Pradhan B, Segoni S, Alamri AM (2020) Rainfall induced landslide studies in Indian Himalayan region: a critical review. Appl Sci 10(7):2466

Eberhardt E (2003) Rock slope stability analysis–utilization of advanced numerical techniques. Earth and Ocean sciences at UBC, 4 pp

EM 1110-2-1902 (2003) Engineering and design-Slope stability. US Army corps of engineering. Engineer Manual, Department of the Army, US Army Corps of Engineers, Washington, DC 20314-1000. https://www.publications.usace.army.mil/Portals/76/Publications/EngineerManuals/EM_1110-2-1902.pdf

Froude MJ, Petley DN (2018) Global fatal landslide occurrence from 2004 to 2016. Nat Hazard 18(8):2161–2181

Gilli JA, Corominas J, Rius J (2000) Using global positioning system technique in landslide monitoring. Eng Geol 55(3):167–192

Highland L, Bobrowsky PT (2008) The landslide handbook: a guide to understanding landslides. US Geological Survey, Reston, p. 129

Hyndman D, Hyndman D (2016) Natural hazards and disasters. Cengage Learning

Indian Standard, I.S. (2002) Indian Standard, Criteria for earthquake resistance design of structures, Fifth Revision, Part-I. Bureau of Indian Standard, New Delhi

ISC catalogue: International Seismological Centre (2020) On-line Bulletin, https://doi.org/10.31905/D808B830, In-text pls refer as (ISC, 2020, last accessed: 29th Dec 2020)

Hibbitt D, Karlsson B, Sorensen P (2004) ABAQUS analysis user's manual. Providence, RI

Jain AK, Manickavavasagam RM, Singh S (2002) Himalayan collision tectonics. Gondwana Research Group Memoir No. 7, 114 pp

Keefer DK (1984) Landslides caused by earthquakes. Geol Soc Am Bull 95(4):406–421

Leech ML, Singh S, Jain AK, Klemperer SL, Manickavasagam RM (2005) The onset of India-Asia continental collision: early, steep subduction required by the timing of UHP metamorphism in the western Himalaya. Earth Planet Sci Lett 234(1–2):83–97

Leech ML, Singh S, Jain AK (2007) Continuous metamorphic zircon growth and interpretation of U-Pb SHRIMP dating: an example from the Western Himalaya. Intern Geol Rev 49:313–328

Linder W (2009) Digital photogrammetry: a ractical course. Springer, Berlin Heidelberg, p 220

Lowrie W (2007) Fundamentals of geophysics. Cambridge University Press, Cambridge, p 381

Manglik A, Adilakshmi L, Suresh M, Thiagarajan S (2015) Thick sedimentary sequence around Bahraich in the northern part of the central Ganga foreland basin. Tectonophysics 653:33–40

Mugnier JL, Huyghe P (2006) Ganges basin geometry records a pre-15 Ma isostatic rebound of Himalaya. Geology 34(6):445–448

Nadim F, Kjekstad O, Peduzzi P, Herold C, Jaedicke C (2006) Global landslide and avalanche hotspots. Landslides 3(2):159–173

NCS catalogue: National Center for Seismology (2020) On-line Bulletin, https://seismo.gov.in/seismological-data. In text pls refer as (NCS, 2020; last accessed: 29th Dec 2020)

Parkash S (2013) Earthquake related landslides in the Indian Himalaya: experiences from the past and implications for the future. In: Landslide science and practice. Springer, Berlin, Heidelberg, pp 327–334
Prasath RA, Paul A, Singh S (2019) Earthquakes in the Garhwal Himalaya of the central seismic gap: a study of historical and present seismicity and their implications to the seismotectonics. Pure Appl Geophys 176(11):4661–4685
Singh IB (1996) Geological evolution of Ganga Plain—an overview. J Palaeontol Soc India 41:99–137
Singh S (2019) Protracted zircon growth in migmatites and in situ melt of Higher Himalayan Crystallines: U–Pb ages from Bhagirathi Valley, NW Himalaya, India. Geosci Front 10(3):793–809
Singh S (2020) Himalayan Magmatism through space and time. Episodes 43(1):358–368
Singh S, Jain AK (2003) Himalayan granitoids. J Virtual Explor 11:1–20
Singh S, Jain AK (2007) Liquefaction and fluidization of lacustrine deposits from Lahaul-Spiti and Ladakh Himalaya: geological evidences of paleoseismicity along active fault zone. Sed Geol 196:47–57
Singh S, Singh M (2020) Spatial variability of Sr isotope of Gomati river basin within Ganga Alluvial Plain: Implications for global seawater fluxioning. Geochem J 54(2):57–70
Shukla UK, Srivastava P, Singh IB (2012) Migration of the Ganga River and development of cliffs in the Varanasi region, India during the late quaternary: role of active tectonics. Geomorphology 171:101–113
Srinivas D, Srinagesh D, Chadha RK, Ravi Kumar M (2013) Sedimentary thickness variation in the indo-gangetic foredeep from inversion of receiver functions. Bull Seismol Soc Am 103(4):2257–2265. https://doi.org/10.1785/0120120046
Su K, Li Y, Cheng D (2016) Slope stability analysis under combined failure criteria. Open Civil Eng J 10(1)
Tiwari A, Narayan AB, Dwivedi R, Dikshit O, Nagarajan B (2020) Monitoring of landslide activity at the Sirobagarh landslide, Uttarakhand, India, using LiDAR, SAR interferometry and geodetic surveys. Geocarto Int 35(5):535–558. https://doi.org/10.1080/10106049.2018.1524516
Thakur VC (1996) Landslide hazard management and control in India. ICIMOD pulication, 51 pp
Varnes DJ (1978) Slope movement types and processes. Special Report 176:11–33
Yang W, Shen L, Shi P (2015) Mapping landslide risk of the world. In: World atlas of natural disaster risk. Springer, Berlin, Heidelberg, pp 57–66
Yin A (2006) Cenozoic tectonic evolution of the Himalayan orogen as constrained by along-strike variation of structural geometry, exhumation history, and foreland sedimentation. Earth Sci Rev 76(1–2):1–131

Landslides in the Himalayas: Causes, Evolution, and Mitigation—A Case Study of National Highway 44, India

Mohsin Fayaz, Sheik Abdul Khader, and Mohammd Rafiq

Abstract A landslide is a natural phenomenon influenced by gravity which causes downward movement of materials such as soil, rocks, mud, and so on. In the past, landslides have triggered numerous mishaps and are a key danger to human life and assets. Due to the collision of Indian and Eurasian plates, Himalayan mountain range was formed after a huge bang. Continuous moment of plates makes the landscape/slope brittle, fragile, and vulnerable to landslides. Risk preparedness and an effective alert system are necessary to avoid the loss of human life and property. There are natural as well as anthropogenic factors that trigger landslides on the steep slopes. Various data analysis operations, remote sensing techniques, and field sampling experiments (Direct share, natural density, Atterberg Limits (plastic limit, liquid limit, plasticity index) Moisture content, and Specific gravity) are performed to find out the leading cause of land failures to mitigate its effects on the local population. This chapter focusses on the causes of the occurrence of the landslides on the national highway NH-44 and concludes with the mitigation and management measures to be undertaken in order to reduce the damages it imbibes on the economy of the UT of Jammu and Kashmir, India.

Keywords Landslide · Disaster · Himalayas · Early warning system · Natural capital

1 Introduction

Landslides can be induced by various factors (Anthropogenic, Morphological, or geological), but the most frequent landslides are caused by prolonged precipitation and anthropogenic activities (Crozier 2010). The Himalayas receive a large

M. Fayaz (✉) · S. A. Khader
Department of Computer Applications, BS. Abdur Rahman Crescent Institute of Science and Technology, Vandalur, Tamil Nadu, India

M. Rafiq
Department of Space, Government of India, National Atmospheric Research Laboratory, Gadanki, Andhra Pradesh, India

S. Kanga et al. (eds.), *Disaster Management in the Complex Himalayan Terrains*, Geography of the Physical Environment,
https://doi.org/10.1007/978-3-030-89308-8_4

amount of rainfall throughout the year, which highly saturates the soil and rocks on the steep slopes and becomes a cause for land failures (Altaf et al. 2013, 2014; Dikshit et al. 2019). There are many landslide-prone spots on the Himalayas, but the Jammu Srinagar National Highway route is considered one of the dangerous roads that connect Kashmir Valley with the rest of the country (Fayaz and Khader 2020). Several landslide-prone zones on the National Highway experience frequent occurrences of landslips caused by heavy precipitation during monsoon and winter seasons. Landslides along National Highway greatly impact the economy of Kashmir valley. According to a report published in a daily newspaper "Greater Kashmir," closure of national highway (21 times) in November and December 2019 has cost the valley economy around 1995 Crore rupees which is a loss of 95 crore rupees per day and has claimed nearly 447 lives due to the accidents in the same year. In this chapter, we will analyze and investigate one of the sites at the national highway. The landslides have caused many casualties and disasters in the Himalayas. On August 2, 2014, the massive Himalayan landslide that struck Nepal's Sindhupal chowk district wiped out an entire village and damaged two others. Another fatal landslide struck Malin' a village in Maharastra, India 30, July 2014.

To identify the cause of the landslide, soil sampling, and testing is carried out using the conventional American Society for Testing and Materials (ASTM) processes. A soil specimen is extracted from the depth of 3 m and is properly sealed to protect the sample from any external climatic conditions, external shaking, or roiling. The specimen is then tested and investigated in the laboratory using various scientific testing methods under different conditions and circumstances. The impact of various soil properties (Atterberg Limits, Direct Share, Moisture Content, and natural density) are evaluated and analyzed for triggering landslides. Remote sensing data like land surface temperature (LST), Digital Elevation Model are used to determine the cause of landslides in the study area. Using Moderate Resolution Imaging Spectroradiometer (MODIS) data, LST from 2003 to 2020 is analyzed to find out the average land surface temperature at the study area. The LST is directly proportional to the groundwater level, which is also one of the leading agents causing landslides (Rafiq et al. 2012).

2 Description of the Study Area

The study area is located (75.11 E, 33.2 N, 75.12 E, 33.21 N) near the Ramban District of Jammu and Kashmir, shown in Fig. 1 is highly prone to landslides and has recorded many landslides in the past. The region has a hilly topography with an average altitude of 2740 m above mean sea level (AMSL) (Fayaz and Khader 2020). Deforestation and anthropogenic activities leading to instability of steep slopes along the National Highway are also responsible for frequent landslides and rockfalls (Kumar et al. 2008; Nathawat et al. 2010; Rather et al. 2017). Steep slopes are also a significant conditioning factor for landslides/rock falls because the slope angle is higher than 15 degrees, which creates increased pressure on soil and material on the slope due to

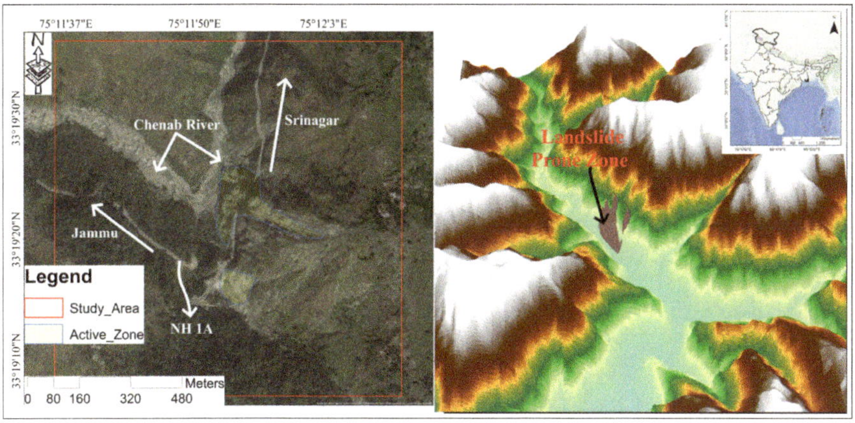

Fig. 1 Study Area and the Topographical setting of active landslide zone

gravity (Booth et al. 1993). The anthropogenic activities like the widening of roads and construction of tunnels in an unscientific manner have made slopes fragile and unstable in the study area (Chingkhei et al. 2013). Steep hill cuttings and vibrations have developed massive fractures on the slopes, resulting in deadly landslides (Ralh and Ralh 2013). Figure 1 shows the study area at the national highway, which is highly prone to landslides.

The study area's steep slope is highly disturbed, which has made the soil loose enough so that it can subsequently fall during heavy precipitation events (Farooq and Muslim 2014; Meraj et al. 2015, 2016). The study area covers an area of 674.45 ft with significantly less vegetation cover over it. Less vegetation at the study area makes it more prone to failures because the vegetation cover plays an influential role in stabilizing soil and is useful in increasing soil stability (Collison and Anderson 1996; Kanga et al. 2017a, b; Kumar et al. 2018a, b). Slope with ideal vegetative cover and root mass will stay up to 5° steeper than bare slope (Julia Wessels 2017).

2.1 Climate and Rainfall

The Himalayas climate has never been uniform as the temperature falls by 0.2–1.2 degrees – Celsius after a rise in every 100 m of altitude (Romshoo et al. 2018, 2020). This gives rise to a range of climate changes from cold and snow-covered mountains at high altitudes and warm at low altitudes (foothills) (Gujree et al. 2017; Kanga et al. 2017a, b). The study area temperature ranges from minus 5 in winters and 20–25 degrees Celsius in the summer season (Fayaz and Khader 2020). Heavy rain arrives in June till September (depending on Monsoon) (Mifatul Shafiq et al. 2019). The heavy precipitation over the study area can cause menacing landslides over a period which can cause massive damage to the transport and local population if no

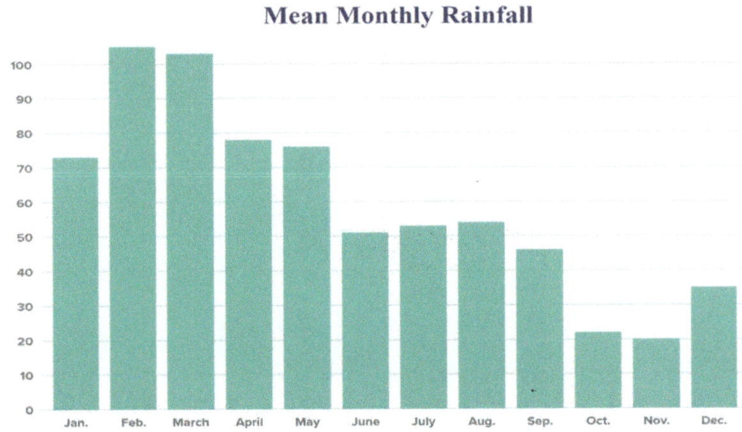

Fig. 2 Mean monthly rainfall of study area from the year 2000–2019

precautions are taken. Due to the western disturbances that travel from west to east, heavy precipitation is recorded over the Himalayas and Trans Himalayas.

In the months with heavy precipitation, many rainfall-induced landslides are recorded on the Jammu Srinagar National Highway. The heavy rainfall oversaturates the disturbed soil and rocks further to make landfall (Kumar et al. 2017). The average monthly rainfall of the area from the year 2000–2020 is shown in Fig. 2. The rainfall data were collected from the nearest IMD station.

The months receiving high rainfall during 2000–2019 are February, March, April, and May, which indicates the increase in the number of landslide risks during these months.

3 Causes of Landslides on the Himalayas

Landslides are triggered by natural and anthropogenic activities such as erosion, geological weathering, volcanic activity, and deforestation (Hasegawa et al. 2009; Singh and Kanga 2017). An increased risk of landslides has been associated with the destruction of vegetation by droughts, fires, and logging (Zheng 2006). As the angle of slope increases (makes it steeper), the parallel component of gravity increases, and the perpendicular component decreases, thereby eliminating resistance to downward motion (Iwahashi et al. 2003). When a similar component is more significant than the vertical part, the object will slip down the slope.

Most of the landslides have occurred on the Himalayas due to heavy rainfall events and human activities like roads over the mountains (Saha et al. 2002). Anthropogenic activities like the widening of tunnels' roads and construction in an unscientific manner have made slopes fragile and unstable (Fayaz and Khader 2020). Steep hill cuttings and vibrations have developed massive fractures on slopes resulting in deadly

Fig. 3 Landslide at study area triggered due to the heavy rainfall during year 2019 (25-02-19)

landslides and rockfalls. Figure 3 below shows a landslide at the study area, which was triggered during a heavy rainfall event.

Soil and rocks have become fragile and loose that even a short spell of rainfall can cause the land to slide down. The leading cause of the landslide soil samples was collected and tested using conventional ASTM processes. These samples were tested to determine the parameters such as soil moisture, natural density, specific gravity, direct share, etc.

3.1 Soil Moisture

Soil moisture is an essential parameter that helps in soil stability and vegetation growth (Veihmeyer and Hendrickson 1950). Excessive moisture can cause the soil to slide over steep slopes due to the loss is share strength (Jotisankasa and Vathananukij 2008). The shear strength of the soil may be a resistance to stress sharing and a consequent susceptibility for shear deformation and is the magnitude of the shear stress a soil can withstand (Godt et al. 2009).

Earth attains the shear strength from:

1. Cohesion ("Adhesion between the soil particles")
2. Internal friction of the soil particles

Rainfall and melting of snow on the mountains are the two main components that increase the soil's moisture content and play the main role in triggering landslides on Himalayas (Mishra and Rafiq 2017). It remains to change with the changes in precipitation and temperature. Excessive soil moisture lowers the angle of internal friction between the soil particles and makes them loose enough to get flown (Ray and Jacobs 2007). The soil sample was tested for soil moisture, and the moisture content of the soil was recorded as 27.33%. The sample was also tested to find out the Atterberg's Limit, which is described as the boundaries of the soil states (Liquid Limit, Plastic Limit, Plasticity Index) when getting saturated with different moisture contents. It can help determine or predict the soil states on the slope when properly monitored with the soil moisture sensors. Liquid limit is the state when the soil is saturated enough to act like liquid or semi-liquid (Sharma and Bora 2003). The liquid limit can be considered as a high-level warning for landslides. The limits obtained are Plastic Limit = 19.43, Liquid Limit = 43.22, Plasticity Index = 23.79.

The plastic limit is the boundary between the solid and plastic state of the soil. This limit can be considered as a moderate level warning for landslides while the liquid limit is the boundary between the plastic and liquid limit. It separates the viscous liquid state from the plastic state of the soil (Guo-bing and Guo-bing 2005). The liquid limit is extreme when the soil starts to move down the slope like the viscous liquid. It can be considered a high-level warning for landslides (Fayaz and Khader 2020). The liquid limit of the soil is the maximum moisture the soil can retain. After attaining this moisture level, the soil starts to slide (Stark et al. 2005). The soil at the study area is disturbed and loose enough to begin to run off after the moisture level reaches the liquid limit. Hence, we can say that the soil's liquid limit is the high-level warning (Red Alert) for the landslide (Fayaz and Khader 2020).

3.2 *Natural Density*

Natural density is defined as the weight of dry soil per unit volume, which is articulated as gm's/cm^3 (Labelle and Jaeger 2011). Natural density, also known as bulk density (BD), helps determine the size, arrangement, and shape of the soil particles and the voids. It is responsible for root growth and soil permeability (Brasher et al. 1966). It is usually beneficial to have a low BD soil (<1.5 g/cm^3) to allow the optimal air and water flow through the soil. Soil with higher BD restricts the root growth, making the soil loose and prone to landslides (Singh and Sainju 1998). Vegetation helps to stabilize and increase the slope stability in various ways like; it helps in root reinforcement, anchored and embedded stalks may serve as slope pillars, or as pillars of arches that mitigate shear stress and modifies (reduces) the soil moisture. The study area's BD was found as 1.91 g/cm^3, which is relatively high and unhealthy for root growth. So, we can consider this as the main reason for less vegetation in the study area.

3.3 Specific Gravity

Specific soil gravity (GS) is the unit weight ratio of solid particles to the unit weight of water at a temperature of 4 °C that usually is between 2.65 and 2.80 (Laboski and Kelling 2007).

$$GS = \frac{\gamma s}{\gamma W}$$

where

$'\gamma S'$ Is the unit weight of the solid particles.

$'\gamma W'$ Is the unit weight of the water at 4 °C.

The lesser value of GS defines the granular particles, while the higher value defines the inorganic clay. The study area's soil sample shows the GS value of 2.84, which is high, so it means the soil at the study area is mostly composed of both inorganic clay and granular particles. Due to the anthropogenic activities and weathering, the slope has become week enough and highly prone to landslides. The heavy precipitation at the study area is believed to not only contribute to the weathering of the rock mass. However, it also increases the soil's moisture content (clay), which reduces the slope's stability.

3.4 Direct Share

Direct share is the test used to find the soil's share strength; share strength helps geotechnical structures from collapsing (Cohen et al. 2009). Failure happens where the shear stress exceeds its limit of share strength. In simple words, it is the maximum share of stress soil can attain. It depends on how the soil particles are interlocked and how much stress these particles can achieve without getting rolled down. Share strength is attained from two sources: cohesion (C) and frictional resistance between the particles (Phi).

Share strength is written (equation) as:

$$S = C' + \delta * \tan \varphi'.$$

Where:

C′ Effective Cohesion
Δ Effective Stress
φ′ Effective Angle of Sharing Stress.

The two main forces which help to attain the shared strength are discussed below:

(a) **Cohesion (C)**:
Cohesion is a significant soil stability element that helps the soil particles to withstand the shear stress (Matsushi et al. 2006). It is the inter-particular attraction between the soil particles which holds them up. Coarse-grained soil is considered the less cohesive soil, while the clay is the better example of highly cohesive soil (Lombardi et al. 2013). Clay has strong cohesion between its particles but has less angle of internal friction so that the shared strength will be contributed only by the cohesion alone.
Share stress on the soil is contributed by the steep slopes and excessive moisture, decreasing the inter-particular attraction of soil particles. The share stress ratio is often stated as the safety factor for slope stability. If the ratio is below 1.0, then there are very high chances of slope failure.

$$\text{Safety factor}\left(S_f\right) = \frac{St}{Ss}$$

where

S_t Share strength
S_s Share stress.

The cohesion value of 0.97 kg/cm^2 was recorded after the soil sample was tested in the laboratory. The value indicates that the study area's soil is acting as medium clay with the coarser-grained particles of metamorphic rocks. It suggests that a prolonged event of precipitation is enough to cause a landslide in the study area.

(b) **Angle of Internal Friction (Phi)**:
A measure of the strength of a rock or soil to withstand shear stress. It is the angle at which a solid or granular substance may be loaded without slipping or collapsing. It is the angle, determined between the resulting force (R) and the normal force (N), its tangent (S/N) is the coefficient of sliding friction. The landslide potential index driven value indicates that areas (slopes) with a friction angle of fewer than 20° are incredibly likely to collapse. The phi value of 17° was found for the study area, which is low and prone to landslide. Soil friction angle of 17 is located in inorganic clays of high plasticity, which has loose particle packing and less particle surface friction. It gets easily flown after getting saturated enough with the prolonged precipitation events.

4 Earthquake

Most of the slope instabilities are caused by moderate to high earthquakes. Due to the ground vibrations and movements, the slope gets disturbed, which creates extra interstices (pores) in the soil, these pores when getting filled with rainfall increases the water pressure in the soil and rocks, which lowers its mechanical strength (share strength). When the soil's share strength decreases, the shared stress acting on it makes the soil slide, which gets formed into a fatal landslide.

It has also been found that Earthquakes can also influence the slope's stability in the long run. Earthquakes can weaken the slopes and make them susceptible to landslides during subsequent precipitation events. Northwestern Himalaya is one of the most tectonically active domains of the Himalayan arc (Malik and Mohanty 2007). The predominant dynamic collisional tectonic setup is capable of causing devastating earthquakes and landslides. The area falls in Seismic zone IV and V (Kamp et al. 2010). So, earthquakes can also be a big reason for landslides on the Himalayas.

5 Land Surface Temperature (LST)

LST is one of the critical parameter used to forecast the landslide potential areas (Rafiq et al. 2012). It is one of the critical parameters used by researchers around the world to estimate possible landslide areas because the land surface temperature is directly related to the groundwater level (Rafiq et al. 2016, Fig. 4). Previous analysis by the researchers has found that most of the landslide-hit sites have recorded a temperature of 24–26 °C (West et al. 2012).

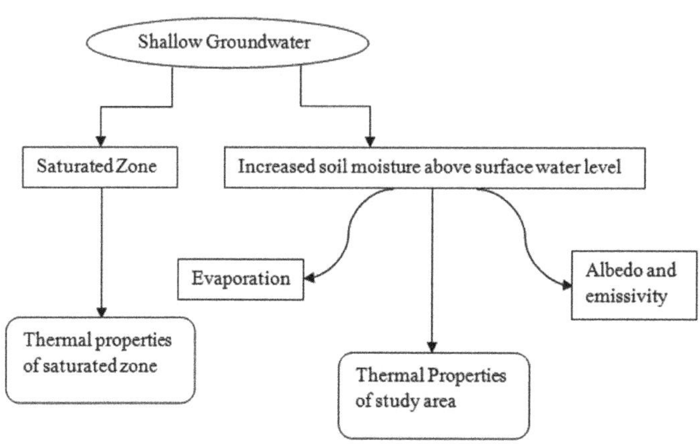

Fig. 4 Schematic overview of groundwater effect

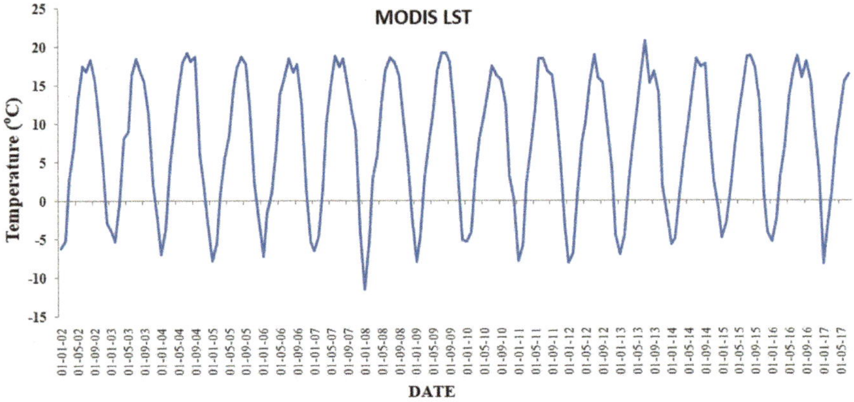

Fig. 5 Land Surface Temperature of the study area

LST is affected by the groundwater, which can be described in two ways.

(a) Direct
(b) Indirect

(a) Direct Effect:
Also referred to as Thermodynamic effect, which is purely based on how the heat gets transferred from one body to another. In this case, the two bodies are land surface and groundwater. These two bodies move thermal energy from hotter to colder. In summer, the ground surface temperature gets transferred to the groundwater, which cools down the land surface. While in winter, the groundwater transfers its heat to the land surface and acts as a heat source (Devaraju et al. 2018).

(b) Indirect:
In this case, the groundwater affects soil moisture (Huang et al. 2007). In some cases, the soil moisture can also help in assuming the groundwater level. The water level can be guessed by measuring the soil moisture.

LST is used to predict the underground water level directly proportional to it to predict the underground water level if we know the LST (Rafiq et al. 2016). The study area's Land surface temperature data in Fig. 5 is generated from the MODIS monthly data product MOD11C3. The study area shows an average land surface temperature of 21.73 °C. Thus, we can say that the study area is highly prone to landslides.

6 High Altitude and Sharp Slopes

We used SRTM DEM data to generate the slope map and the study area's contour map using ArcGIS software. Contour ranges from 1150 to 1450 m (Fig. 6a) with a

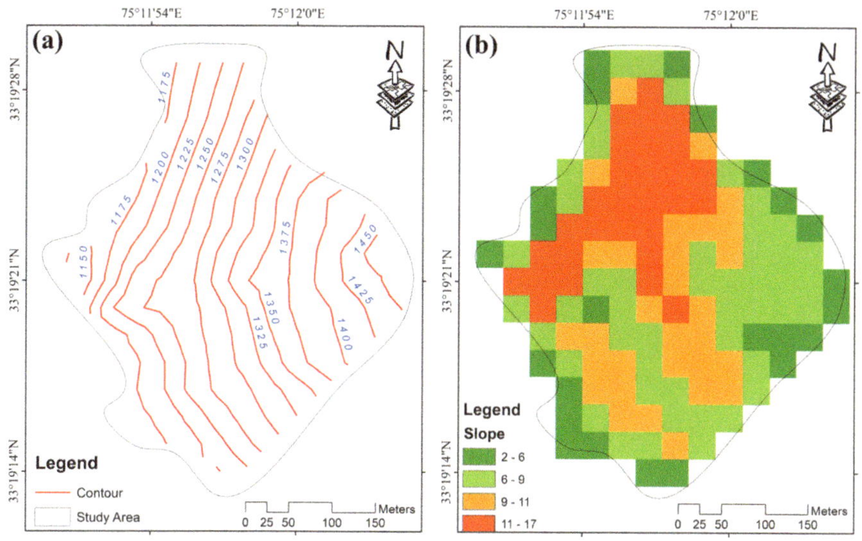

Fig. 6 **a** Contour map and **b** slope map of the active landslide zone in the study area

very steep slope (Fig. 6b), which can influence a landslide. The study area is thus prone to landslides as the slope fall above 80° mostly.

7 Evolution

In this study, we mainly focus on a single active landslide with 674.45 ft. The evolution of this landslide has been monitored for the last two years. Various activities were analyzed and observed in the study area during this period. Activities which we can say are the reason for the evolution of landslide at the study are as under;

(a) Anthropogenic activity (widening of the national highway)
(b) Heavy vehicular traffic
(c) Heavy precipitation

(a) **Anthropogenic Activity**:

Anthropogenic Activities (Man-Made Activities) like the construction of roads, road widening, and deforestation in an unscientific manner are the activities that create pressure on the slope and lead to instability of the hill. Steep slope cutting for road widening using heavy machines creates heavy vibrations, resulting in soil loosening and rocks over the slope. Over the past years, the construction of tunnels and roads has made many slopes fragile/unstable (Siddique et al. 2017), so the number of landslide events will increase drastically shortly.

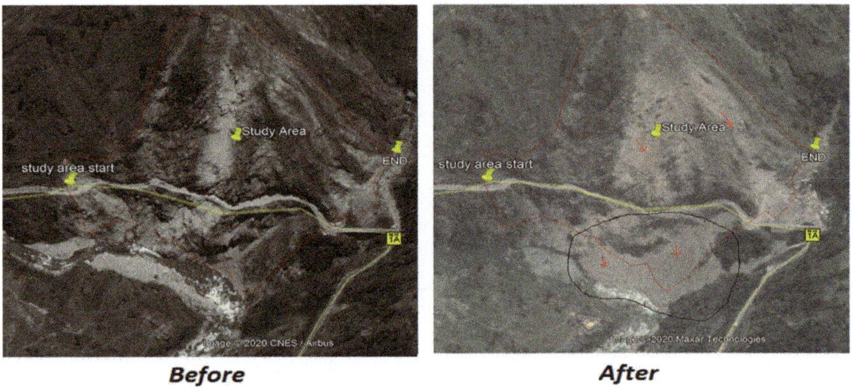

Fig. 7 Study area before and after road widening

Figure 7 shows the changes in National Highway's slope before the road construction (widening) and the present situation. The red arrows indicate the direction of the slide, while the area encircled in black shows the soil, which has a slide from the slope.

(b) **Heavy Vehicular Traffic**:
Heavy vehicular traffic at the Jammu Srinagar National Highway creates heavy vibrations, leading to slope instability and loosening of the soil. Ground vibrations can lead to landslides because the continuous vibration near the steep slope can increase the shared stress on the slope, increasing the soil pore pressure and decreasing the soil's share strength. Thus, the slope becomes extremely vulnerable to collapse due to a decrease in share strength of the soil.

(c) **Heavy Precipitation**:
Prolonged or heavy rainfall is the most common cause of landslides on the Himalayas. Excessive soil moisture lowers the cohesion as well as the angle of internal friction between the soil particles and makes it loose enough to get flown from the slop. The Himalayas are tectonically active with steep slopes, so high precipitation during Monsoon yields an increased risk of landslides.

8 Mitigation

Landslide mitigation includes various methods carried out on the slopes to minimize the impact of slides. Landslides may be triggered by multiple anthropogenic and natural activities like deforestation, heavy rainfall, earthquakes, and cutting slopes in an unscientific way. The landslide risk mitigation is usually not categorized according to the phenomenon that causes landslides, but there are various direct methods used for landslide risk mitigation:

(a) **Early Warning System**:

There are many factors which induce landslides such as heavy precipitation, floods, and earthquake. For avoiding the brunt of landslides, it is crucial to develop a highly sensitive Early Warning System (EWS), which can help detect landslides using various types of forecast models, Moisture balance model, Susceptibility zonation mapping, and rainfall threshold. It is essential to develop a sophisticated LEWS so that the landslides can be forecasted early to allow the local community to take mitigation actions.

(b) **Construction of Retaining Walls (Bunds)**:

It is vital to construct bunds at the landslide-prone areas to stop the slide; it can also fix the slope instability by supporting the fractured or sharp slopes.

(c) **Afforestation**:

Afforestation on the steep and disturbing slopes helps the slope to regain strength by reattaching the soil particles together firmly. It reduces soil erosion by restricting excessive rainfall (precipitation) into the soil. Vegetation holds the soil particles and binds them together, which provides excellent strength to the slope by reinforcing soil shear strength. Conversion of vegetation and range improvement activities on steep slopes may dramatically impact the site stability. Proper vegetation management on the steep slopes helps the slope to regain composure and strength (Kumar and Bhagavanulu 2008).

(d) **Restricting Construction at Prone Sites**:

The district disaster management authority should ban constructions near landslide-prone sites because it can be dangerous and life-threatening for the local population.

(e) **Drainage**:

Deep or shallow drainage at the site can increase the slope's stability by decreasing water volume within a vulnerable pitch. Depending on the slope morphology, the deep or shallow drain can reduce the pore pressure and excessive soil moisture of the soil, strengthening the soil on the slope and minimizing the chances of slope failure. Shallow drainage is used to drain out the surface water, while deep drainage is used to drain out the water inside the slope. Using both types of drainage is much effective and safe.

9 The Way Forward

Landslides along National Highway are one of the most critical hazards. Hence, finding out the main triggering factor is essential for adequately monitoring prone zones located on National Highway. The vital elements that trigger landslides include prolonged or intensive precipitation events, sharp slopes, ground water level changes, increased soil moisture, and heavy traffic near the prone area. The proper drainage system is essential to reduce the rainwater infiltration, which can reduce the soil moisture of the slope. Soil Moisture is an important parameter that helps in soil stability and vegetation growth (Veihmeyer and Hendrickson 1950; Meraj et al. 2013). Excessive moisture can cause the soil to slide over the slopes due to the loss in its share strength (Jotisankasa and Vathananukij 2008). Shallow and deep drainage should be used to drain out the extra water from the slope, which is the best way to reduce the slope's pore water pressure. Early Warning System for such disasters can help to save precious lives and properties. The existing (Machine Learning) models' prediction accuracy needs to be improved for more accurate and precise warnings. The model can be improved by including new, responsible parameters (variables) to enhance prediction accuracy. Human casualty and property loss can be reduced by restricting all constructions and installing the piles and retaining walls near the highly prone landslide zones.

References

Altaf S, Meraj G, Romshoo SA (2014) Morphometry and land cover based multi-criteria analysis for assessing the soil erosion susceptibility of the western Himalayan watershed. Environ Monit Assess 186(12):8391–8412

Booth JS, O'leary DW, Popenoe P, Danforth WW, Schwab WC (1993) US Atlantic continental slope landslides: their distribution, general attributes, and implications. In: Schwab WC, Lee HJ, Twichell DC (eds) Submarine landslides: selected studies in the US Exclusive Economic Zone, pp 14–22

Brasher BR, Franzmeier DP, Valassis V, Davidson SE (1966) Use of saran resin to coat natural soil clods for bulk-density and water-retention measurements. Soil Sci 101(2):108

Chingkhei RK, Shiroyleima A, Singh LR, Kumar A (2013) Landslide hazard zonation in NH-1A in Kashmir Himalaya, India. Int J Geosci

Cohen D, Lehmann P, Or D (2009) Fiber bundle model for multiscale modeling of hydromechanical triggering of shallow landslides. Water Resources Res 45(10)

Collison AJC, Anderson MG (1996) Using a combined slope hydrology/stability model to identify suitable conditions for landslide prevention by vegetation in the humid tropics. Earth Surf Proc Land 21(8):737–747

Crozier MJ (2010) Deciphering the effect of climate change on landslide activity: a review. Geomorphology 124(3–4):260–267

Devaraju N, de Noblet-Ducoudré N, Quesada B, Bala G (2018) Quantifying the relative importance of direct and indirect biophysical effects of deforestation on surface temperature and teleconnections. J Clim 31(10):3811–3829

Dikshit A, Sarkar R, Pradhan B, Acharya S, Dorji K (2019) Estimating rainfall thresholds for landslide occurrence in the Bhutan Himalayas. Water 11(8):1616

Farooq M, Muslim M (2014) Dynamics and forecasting of population growth and urban expansion in Srinagar city-a geospatial approach. Int Arch Photogrammetry Remote Sens Spat Inf Sci 40(8):709

Fayaz M, Khader SA (2020) Identifying the parameters responsible for Landslides on NH-44 Jammu Srinagar National Highway for Early Warning System. Disaster Advances 13(2)

Godt JW, Baum RL, Lu N (2009) Landsliding in partially saturated materials. Geophys Res Lett 36(2)

Gujree I et al (2017) Evaluating the variability and trends in extreme climate events in the Kashmir valley using PRECIS RCM simulations. Model Earth Syst Environm 3(4):1647–1662

Guo-bing Z, Gai-ning YU (2005) Study on improvement of high liquid limit soil. J Highway and Transp Res Dev 22:26–29

Hasegawa S, Dahal RK, Yamanaka M, Bhandary NP, Yatabe R, Inagaki H (2009) Causes of large-scale landslides in the Lesser Himalaya of central Nepal. Environ Geol 57(6):1423–1434

Huang Y, Chameides WL, Dickinson RE (2007) Direct and indirect effects of anthropogenic aerosols on regional precipitation over east Asia. J Geophys Res Atmos 112(D3)

Iwahashi J, Watanabe S, Furuya T (2003) Mean slope-angle frequency distribution and size frequency distribution of landslide masses in Higashikubiki area, Japan. Geomorphology 50(4):349–364

Jotisankasa A, Vathananukij H (2008, November) Investigation of soil moisture characteristics of landslide-prone slopes in Thailand. In: International conference on management of landslide hazard in the Asia-Pacific Region 11th–15th November, p 12

Kamp U, Owen LA, Growley BJ, Khattak GA (2010) Back analysis of landslide susceptibility zonation mapping for the 2005 Kashmir earthquake: an assessment of the reliability of susceptibility zoning maps. Nat Hazards 54(1):1–25

Kanga S, Kumar S, Singh SK (2017a) Climate induced variation in forest fire using remote sensing and GIS in Bilaspur district of Himachal Pradesh. Int J Eng Comput Sci 6(6):21,695–21,702

Kanga S, Singh SK, Sudhanshu (2017b) Delineation of urban built-up and change detection analysis using multi-temporal satellite images. Int J Rec Res Aspects 4(3):1–9

Kumar A, Asthana AKL, Priyanka RS, Jayangondaperumal R, Gupta AK, Bhakuni SS (2017) Assessment of landslide hazards induced by extreme rainfall event in Jammu and Kashmir Himalaya, northwest India. Geomorphology 284:72–87

Kumar SV, Bhagavanulu DVS (2008) Effect of deforestation on landslides in Nilgiris district—a case study. J Indian Soc Remote Sens 36(1):105

Kumar S, Singh SK, Mahendra RS (2018a) Assessment of land use and land cover during pre and post cyclone Phailin in southern part of Odisha coast. i-Manager's J Future Eng Technol 13(4):53

Kumar S, Kanga S, Singh SK, Mahendra RS (2018) Delineation of Shoreline Change along Chilika Lagoon (Odisha), East Coast of India using Geospatial technique. 2018b. Water Quality Zoning Using GIS & Remote Sensing, 207

Labelle ER, Jaeger D (2011) Soil compaction caused by cut-to-length forest operations and possible short-term natural rehabilitation of soil density. Soil Sci Soc Am J 75(6):2314–2329

Laboski CA, Kelling KA (2007) Influence of fertilizer management and soil fertility on tuber specific gravity: a review. Am J Potato Res 84(4):283–290

Lombardi D, Bhattacharya S, Wood DM (2013) Dynamic soil–structure interaction of monopile supported wind turbines in cohesive soil. Soil Dyn Earthq Eng 49:165–180

Malik JN, Mohanty C (2007) Active tectonic influence on the evolution of drainage and landscape: geomorphic signatures from frontal and hinterland areas along the Northwestern Himalaya, India. J Asian Earth Sci 29(5–6):604–618

Matsushi Y, Hattanji T, Matsukura Y (2006) Mechanisms of shallow landslides on soil-mantled hill-slopes with permeable and impermeable bedrocks in the Boso Peninsula, Japan. Geomorphology 76(1–2):92–108

Meraj G, Yousuf AR, Romshoo SA (2013) Impacts of the geo-environmental setting on the flood vulnerability at watershed scale in the Jhelum basin. MPhil dissertation (2013)

Meraj G et al (2015) Assessing the influence of watershed characteristics on the flood vulnerability of Jhelum basin in Kashmir Himalaya. Nat Hazards 77(1):153–175

Meraj, G, Romshoo SA, Altaf S (2016) Inferring land surface processes from watershed characterization. In: Geostatistical and Geospatial Approaches for the Characterization of Natural Resources in the Environment. Springer, Cham, pp 741–744

Mifatul Shafiq M, Ramzan S, Ahmed P, Mahmood R, Dimri AP (2019) Assessment of present and future climate change over Kashmir Himalayas, India. Theoretical Appl Climatol 137(3–4):3183–3195

Mishra AK, Rafiq M (2017) Analyzing snowfall variability over two locations in Kashmir, India in the context of warming climate. Dyn Atmos Oceans 79:1–9

Nathawat MS et al (2010) Monitoring & analysis of wastelands and its dynamics using multiresolution and temporal satellite data in part of Indian state of Bihar. Int J Geomatics Geosci 1(3):297–307
Rafiq M, Rashid I, Romshoo SA (2012) Estimation and validation of remotely sensed land surface temperature in Kashmir valley. J Himalayan Ecol Sustain Dev 9:2014
Rafiq M, Rashid I, Romshoo SA (2016) Estimating land surface temperature and its lapse rate over Kashmir Valley Using MODIS Data. In: Geostatistical and geospatial approaches for the characterization of natural resources in the environment. Springer, Cham, pp 723–728
Ralh KD, Ralh AD (2013) Landslide at railway project site in weak Murree formation—case study. International conference on case histories in geotechnical engineering
Rather, MA et al (2017) Assessing the influence of watershed characteristics on soil erosion susceptibility of Jhelum basin in Kashmir Himalayas. Arab J Geosci 10(3):59.
Ray RL, Jacobs JM (2007) Relationships among remotely sensed soil moisture, precipitation and landslide events. Nat Hazards 43(2):211–222
Romshoo SA, Rafiq M, Rashid I (2018) Spatio-temporal variation of land surface temperature and temperature lapse rate over mountainous Kashmir Himalaya. J Mt Sci 15(3):563–576
Romshoo SA, Fayaz M, Meraj G, Bahuguna IM (2020) Satellite-observed glacier recession in the Kashmir Himalaya, India, from 1980 to 2018. Environ Monit Assess 192(9):1–17
Saha AK, Gupta RP, Arora MK (2002) GIS-based landslide hazard zonation in the Bhagirathi (Ganga) valley, Himalayas. Int J Remote Sens 23(2):357–369
Sharma B, Bora PK (2003) Plastic limit, liquid limit and undrained shear strength of soil—reappraisal. J Geotech Geoenviron Eng 129(8):774–777
Siddique T, Pradhan SP, Vishal V, Mondal MEA, Singh TN (2017) Stability assessment of Himalayan road cut slopes along National Highway 58, India. Environ Earth Sci 76(22):759
Singh BP, Sainju UM (1998) Soil physical and morphological properties and root growth. HortScience 33(6):966–971
Singh SK, Kanga S (2017) Role of geoinformatics in site suitability analysis of infrastructures using PRA Approach. Am Int J Res Sci Technol Eng Math 1(18):81–85
Singh SK, Mishra SK, Kanga S (2017) Delineation of groundwater potential zone using geospatial techniques for Shimla city, Himachal Pradesh (India). Int J Sci Res Dev 5(4):225–234
Stark TD, Choi H, McCone S (2005) Drained shear strength parameters for analysis of landslides. J Geotech Geoenviron Eng 131(5):575–588
Veihmeyer FJ, Hendrickson AH (1950) Soil moisture in relation to plant growth. Annu Rev Plant Physiol 1(1):285–304
Wessels J, Swiss Federal Research Institute WSL (2017). https://www.waldwissen.net/en/forest-ecology/protective-function/soil-and-rock-slide/plant-diversity-protects-against-landslides
West DC, Shugart HH, Botkin DF (eds) (2012) Forest succession: concepts and application. Springer Science & Business Media
Zheng FL (2006) Effect of vegetation changes on soil erosion on the Loess Plateau. Pedosphere 16(4):420–427

Geo-Environmental Impact of Road Widening Project Along National Highway-44, Jammu and Kashmir, India

Vinay Kumar Pandey

Abstract Himalaya has a complex ecosystem and rugged topography. To connect people in this region by road with the aim of harsh-free travel, the Government of India has started various road projects all over the country. The widening of National Highway-44 is a special project for Jammu and Kashmir given connectivity, transportation, defense security, and tourism. The widening of NH-44 is full of geo-environmental challenges as it crossed Outer Himalaya to Greater Himalaya, having different topography, variable hill slope, various rock conditions, creating difficulties during construction. Climatic and hydro-meteorological factors are playing as a catalyst in construction complications. Implementation of stepwise preventive measures will reduce the risk of landslide. In this chapter, we discuss geo-environmental challenges the construction agencies face, such as landslides and associated impact on the old landslide, failure of residential and agricultural land along the hill slope. Also, we discuss the implications for the foundation of high-tension transmission towers within the selected study area Udhampur–Chenani road section, part of NH-44, Jammu–Srinagar National Highway-44. Based on the types of landslides and related impacts in the Udhampur–Chenani road section, it is observed that area is covered with thick debris/colluvium materials with steep slopes and water bodies (rainfall, nallas, and rivers). To protect houses and transmission towers, the construction agency took modern engineering's help after considering the site feasibility. It proposed a cladding wall, retaining wall, micro-piling at the tower foundation with a waler beam, anchor, and shotcrete. Due to administerial hurdles, it is suggested that an independent Hill Development Authority may be constituted and mandated for preparing long-term plans for hill development in this part of Himalaya after taking care of local geological challenges. Landslide study during the project feasibility stage is highly recommended with mitigation measures that would save the project completion time as well as project cost escalation.

Keywords Geo-environmental challenges · Road widening · Landslide · Slope failure · Precautionary measures

---

V. K. Pandey (✉)
Geological & Geotechnical, Geosystems Infrastructure, IndiaMumbai

S. Kanga et al. (eds.), *Disaster Management in the Complex Himalayan Terrains*, Geography of the Physical Environment,
https://doi.org/10.1007/978-3-030-89308-8_5

1 Introduction

Himalaya is having a complex ecosystem due to its rugged topography, surface and subsurface strata, and living standards of the populations living in them. Topographically, Himalaya is having a very challenging environment according to altitude and moves toward the northside of the Himalaya. Hillslope, geology, vegetation cover, hydrology, and natural disaster are major factors that are affecting the Himalayan ecosystem (Nathawat et al. 2010; Meraj et al. 2015, 2016). For connectivity of local public, government, as well as local public, had created the foot track, village road, motor-able road according to hill slope and hazard-free area in the earlier time. Due to industrialization and mechanized era, vehicle and public movements are increasing day by day, and for developing the connectivity and smooth travel in a hilly area, the Government of India has started various projects in Himalaya. National Highway-44 (NH-44, earlier known as NH-1A) is the lifeline of Jammu and Kashmir as well as the Leh–Ladakh area. Distance between Jammu to Quazigund (near Jawahar Tunnel) is 193 km, but the normal travel time is about 8–9 h due to the narrow and zig-zag road. Topographical areas have highly variable; road elevation varies from mean sea level (MSL) 327 m–2400 m–1647 m–2194 m–1670 m (Jammu-Patnitop-Banihal-Jawahar Tunnel-Quazigund). Road blockage due to landslide and snowfall is a normal event along the highway. Climatical area belongs to the humid subtropical region of Jammu to a temperate region of Banihal. The area receives precipitation twice a year from January to March and very heavy rainfall between July and September. Average precipitation from Jammu to Qazigund is 1500–600 mm. Two major rivers are flowing along this road, first Tawi (Jammu to Chenani) and second Chenab River (Peera to Ramban). Many small tributaries were also crossed by NH-44.

2 Geo-Environmental Setup Along NH-44

National Highway-44 is part of Kashmir Himalaya, divided into the Siwalik Himalaya, the Middle Himalaya, and the Greater Himalaya from south to north. The Siwalik Himalaya is the foothills of the Himalayan Range. This region forms a series of parallel narrow rims of the Tertiate Rocks and is structurally bounded in the North and Main/Himalayan Frontal Thrust in the south by the Main Boundary Thrust (MBT) (MFT/HFT). The Main Boundary Thrust (MBT), Panjal Thrust (PT), or Main Central Thrust (MCT), and Zanskar Thrust are the most prominent (Alam et al. 2015a, b; Bhat et al. 2017).

The Siwalik Himalayan comprises of the Siwalik and Murree formation and constitutes semi-consolidated to consolidated sandstones, siltstone, mudstone, shales, conglomerates, and clay beds (Shanker et al. 1989). Structurally, these are folded rocks (Alam et al. 2015a, b; Bhat et al. 2017).

Many series of thrusts are characteristic tectonic features in the Siwalik Himalaya. The most prominent among them are Main Boundary Fault (MBF) which delimits the Murree from the Siwalik strata.

The Middle Himalayan belt is tectonically bounded between the Murree thrust and Panjal Thrust from south to north side (Bhat et al. 2017, Ahmad et al. 2015). The Middle Himalaya is composed of Late-Proterozoic metamorphosed rocks and unfossiliferous to fossiliferous Lower Paleozoic rocks showing mainly of slates, quartzites, volcanics, phyllites, and subordinate metamorphic and sedimentary strata (Shanker et al. 1989) in the sector between Peera to Banihal. Most rock types have a NW–SE to WNW-ESE regional strike with moderate to steep dips north and south (Bhat et al. 2017; Ahmad et al. 2015b).

The Greater Himalayas physiographically depict the highest mountains and include low- to high-quality metamorphic rocks such as shistle and gneisses and complete parorthometamorphite sequence with ignorant proterozoal and younger intrusions.

In view of seismotectonic, earthquake, micro-seismic, and longitudinal and transverse thrusts in the region, the study area has been concluded to be very vulnerable to earthquakes. The partial Jammu and Cashmir is part of Seismic Field IV and partly of Zone V.

3 Geo-Environmental Challenges Along NH-44

Before proposing the project, initial study comprises of geological, geotechnical, and geomechanical properties, and impact on environment of project on area is very essential. Accordingly, project feasibility report has been proposed. We are aware that Himalaya is tectonic active young mountain belt and full of surprises due to variable environment, hydrology, and geological factors. In Himalayan project, there are always some challenging issues which made us more aware for the future project.

3.1 Pre-investigations for Widening of NH-44

To avoid the road curves and shorten the traffic time on NH-44; studies of primary route interpretation of topography, geology, remote sensing, aerial imagery and finalized road alignment with tunnel and bridges is a pre-requisite. In the design and protection of slope, bridge foundation, and geotechnical investigation, a geological mapping along the project alignment provided a support for the subsurface layers along the tunnel. Geological and meteorological hydrological studies have been conducted in order to mitigate natural disasters such as slides, floods, clouds. It was also preferred to study rock burst as required for tunnel construction.

In order to obtain the engineering properties and actual subsurface layers before construction starts, geological field maps as well as various bore holes and geotechnical research were performed. The platform load test/foot load test has been done to verify the in situ bearing capacity of the bridge foundation. In addition, hydraulic fracture testing, a good man jack test, pressure meter testing, permeability testing, geophysical tests on tunnel lines were performed. On the core samples and field soil and rock samples performed for better understanding of the surface profile, laboratory tests, such as the physical properties of materials, unconfined compression strength tests, point loads, triaxial tests, and permeability tests, were performed. The properties of rock engineering and in situ joint orientation study also considered to prevent geological challenges in the preparation of geotechnical and geological reports.

3.2 Types of Geo-Environmental Challenges Along NH-44

NH-44 is passing through highly variable topography, variable hill slope, various rock conditions; creating construction problems. Slope failure, landslide, and associated geo-environmental hazards are main challenges during construction and widening of road. Landslide and associated geological problems during construction of NH-44 are broadly affected through following three factors as mentioned in Table 1.

Table 1 Landslide affecting factors along NH-44 in Himalaya

Geological factors	Hydrological and physical factors	Human factors
• Jointed rock • Adverse orientation of discontinuity • Weathering • Permeability contrast • Slope angle • Earthquake • Fault/thrust movement	• Rainfall • Glacial lake outburst flood (GLOF) • Flood/flash flood • Erosion • Change in vegetation pattern • Loose overburden	• Deforestation • Deep excavation • Mining • Land use change • Vibration/blast loading • Development activities • Reservoirs • Limited ROW for hill cuts

4 Landslide and Associated Impacts Along the Part of NH-44, Udhampur to Chenani Road Section: A Case Study

We are discussing here the landslide and associated impacts faced by civil agency during the road widening along NH-44, and Udhampur–Chenani road section (road chainage Ch 67.00 km to 89.00 km—from south to north direction) is part of Jammu–Srinagar National Highway-44 which is 22 km long, and widening work was awarded in year 2015.

Geologically, the area of study is part of the external range of the Himalayan, consisting of the Murree formation that overlaps the buried surface on the Indian shield's north frontier. These sediments are lithologically semi-consolidated into sandstones, siltstones, mudstones, shale-shaped sandstones, conglomerates, and clay beds. These rocks have wide anticlines and synclines. Structurally, in a nearby area of study, a series of thrust is a distinctive tectonic feature. MBF is near the study area, the most prominent MBF.

The geomorphology of the road section is characterized by moderate to steep valleys. Presence of narrow river valley and numerous small perpendicular tributaries, mild to steep hill slopes, and colluvium deposits at the toe of the slopes are the main geomorphologic features of the project area (Altaf et al. 2014; Rather et al. 2017). The slopes on both sides of river valley are varying generally from 30° to 70°. The geomorphology of area seems highly erosion prone due to presence of weathered rock mass on the slopes, loose rock blocks, old landslide debris, and rock slide debris deposits all along the slopes. The major valleys are being longitudinal and are aligned in WNW-ESE direction.

The study is found to be extremely vulnerable to terramorphosis, earthquake history, micro-seismicity, and longitudinal and transverse thrust problems in the area. The partial Jammu and Kashmir is part of Seismic Field IV and partly of Zone V.

As per approved design and provided right to work (ROW), civil agency had started the hill slope excavation, but due to poor rock/overburden, various slope and ground failure happened and created the unidentified problems such as landslide and its associated impacts, which will be discussed under discussion point of this chapter. The civil agency have faced losses in the form of man power, machineries, precious project cost, and time. Challenges due to unidentified landslide/ground failure issue have been discussed in this case study with probable causes and mitigation measures, which will be beneficial in other project also. Main landslide areas are Bali landslide, Samroli landslide, and Narsoo landslide, which were observed by various researchers such as Verma 1966–67, Ashraf and Nag 1966, Bhat et al. 2002, Singh 2006, Thakur et al. 2010, Chingkhei et al. 2013, Pandey 2018. Major slides are formed along the syncline as mudstone/siltstone is inter-bedded with sandstone with high-angle slope and highly joined rock, and weathered rock and underground water are main causes of landslide.

4.1 Discussions

During the project feasibility study stage, government agencies have proposed hill slope protection work at identified landslide zones such as breast wall, retaining wall, concrete cladding, and other modern engineering solutions. Minor soil failures were also taken care, but during the construction activities, slide zone was activated due to natural slope failure and geological and hydrological factors. In the study area, various sudden slope/ground failure and associated issues have been faced during construction activities which were not identified earlier in the form of activation of old landslide zone, damages of houses, agriculture and forest slope ground, high-tension tower collapse, or in critical condition. Due to these unpredicted ground failures, following adverse impacts have been observed along the study area:

(I) Reactivation of old landslide zones
(II) Agriculture and residential ground slope failure at higher altitude
(III) Failure of foundation of transmission line towers

4.1.1 Reactivation of Old Landslide Zones

During the hill slope excavation at old landslide zones, namely: Bali, Samroli, and Chenani landslides are activated which were covered with thick colluvium material and formed synclinal structure with poor rock at shallow/deep depth. These slides are shown the wedge and planner failures due to an increase of pore water pressure and sometimes undercutting/toe erosion by rivers/water bodies which are activated during rainy season.

The topographical slope in the study area is having 40°–50° with seasonal and perennial water bodies. Physical properties of soil is silty sand in nature with 20–25% water content, 25–41% liquid limit, 17–20% plastic limit as per geotechnical properties (Meraj et al. 2018). Following landslide failures have been faced by civil agency:

(a) (Bali landslide zone: Bali landslide zone is made up of colluvium material with big boulders which came down after the hill excavation. Expected depth of bed rock varies from 15–25 m. This is very big landslide which blocked the road frequently. Sometimes, big boulder is fallen down, and civil agency used the blast to remove the boulder. This slide is so frequent that the civil agency has permanently deputed the machineries for 3–4 months during monsoon season. Photograph of Bali slide is given in Photograph 1, and typical geological cross-sections are shown in Fig. 1. Hindrance of this area is thick colluvium cover, deep poor rock, limited ROW which were not sufficient to maintain the hill slope in gentle angle.

Photograph 1 Site photograph of Bali slide

Photograph 2 Site photograph of Bali slide

(b) Samroli landslide zone: This slide is having differential and aggressive erosion of the mudstone/claystone layer in between the sandstone layers. Soil properties are same as the Bali landslide zone. Expected depth of bed rock varies from 5–30 m. Anticline and syncline geological features are well-developed in between the sandstones and mudstone layers. The crested part of this syncline is dislocated along right limb by a dip-slip fault and along left limb by a tear fault. The sandstone bands are fractured and blocky in nature, and due to the joint orientation, a series of triangular troughs of erosion has been created in the weaker rock. During rainy season, the strata is water charged and lubricated along the joint planes and wedge failure and toe erosion beneath the sandstone band; shown is Samroli slide.

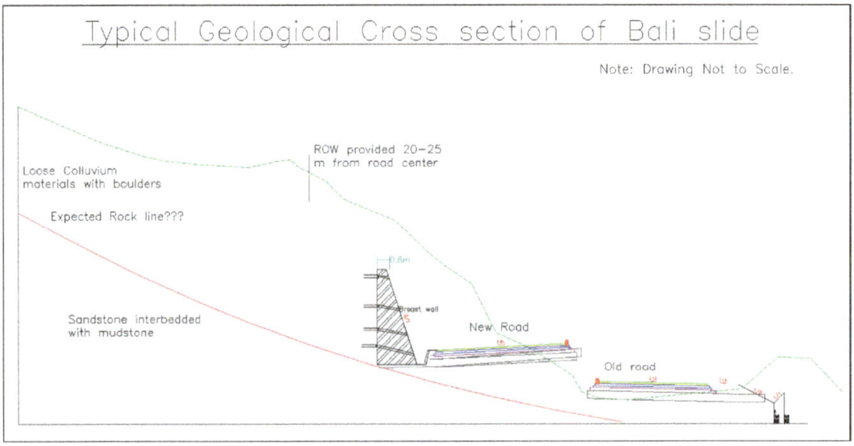

Fig. 1 Typical geological cross-section of Bali slide

The bridge abutment in this location was constructed in the month of August 2016, but in same year in month of September, sliding started and completely blocked NH-44 for one week. Bridge was also covered with sliding mass, about 200 m long and 70–80 m high. More cracks were observed about 60–70 m away from the top of slide area (Photograph 3). The cracks width was 20–30 cm, 50–60 m long, and depth up to 4–5 m and beyond (Photographs 4 and 5). Cracks were developed in two high-tension electricity tower foundation and agriculture land at the top of landslide. Ground water seepage was also observed from landslide. Minor bridge abutment was cracked and settled due to this slide, which completely blocked NH-44 for single lane also (Photograph Fig. 6). The cracks were filled with kanker, sand, cement aggregate. Drainage was also provided to avoid the surface water percolation inside the cracks. Typical geological cross-section is shown as Fig. 2 and shown the impact of landslide after the landslide event happened in September 2016.

Photograph 3 Samroli landslide with dimension and cracks marked, during Sept. 2016 event

Photographs 4 Cracks developed at top of Samroli slide during September 2016 event

Photographs 5 Cracks developed at top of Samroli slide during September 2016 event

Photograph 6 Settlement of bridge and cracks in bridge abutment

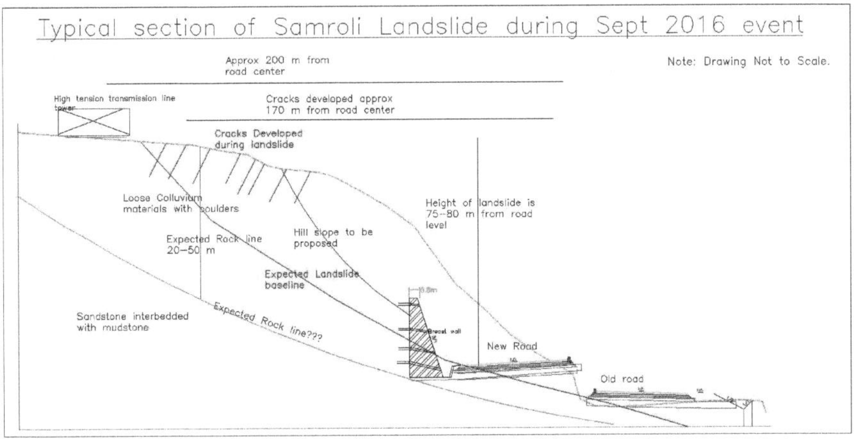

Fig. 2 Typical section of Samroli landslide during Sept. 2016 event

(c) Chenani landslide zone: Chenani landslide zone has colluvium material with 12–18% water content, 25–41% liquid limit, 17–20% plastic limit, having silty sand (SM) in nature above rock. This area is about 150 m wide along the existing road level. It is largely a combination of slump and flow of surface material. The slide material is composed of mainly rock fragments in a matrix of brownish gray silty sand. The maximum thickness of the colluvium is more than 10–20 m. The existing road had been affected by the slide several times. This slide was activated during the monsoon, due to abrupt floods.

4.2 *Agriculture and Residential Ground Failure at Higher Altitude*

Along the hill slope, the houses and agriculture land were developed on colluvium overburden after the deforestation. Due to road widening activities, hill slope was disturbed and sliding started which causes the development of cracks in agriculture and residential ground above and higher altitude of road cutting area. The civil agency had observed grounds sinking at higher altitude at Bali village, Todli village, and Narsoo village.

Maximum numbers of houses were damage reported in the Bali landslide zone. Houses were near the provided ROW and situated on thick colluvium materials—silty soil with big boulders, and slope is gentle with medium cover of vegetation. Rock present at deep level and slope angle is also high. Geomorphological, hydrological, and geological areas are prone to landslide. As bounding strength of colluvium materials is very less and due to water percolation in subsurface, lubrication form at contact plane of rock and colluvium are the favorable factors for landslide as shown in Photographs 7 and 8.

Photograph 7 House damage and geological strata near the ROW at Bali village

Photograph 8 House damage and geological strata near the ROW at Bali village

Next, most ground failures happened at Todli village where houses were very near the ROW of project. Geological area is made of thin colluvium materials 3–4 m before the siltstone inter-bedded with sandstone rock, dipping gentle toward valley side, refer Photograph 9.

The Narsoo village was also affected due to ground settlement and house damage. Normally, houses are away from ROW (100–150 m). But due to poor geological setup of area, thick colluvium materials (3–4 m) and siltstone inter-bedded with sandstone rock, dipping gentle toward valley side, favor the landslide after the starting of road work referred in Photograph 10.

Photograph 9 House damage, tower tilt, and geological strata near the ROW at Todli village

Photograph 10 House damage in critical condition and protection work at Todli village

4.3 *Failure of Foundation of High-Tension Transmission Line Towers*

Protection of high-tension transmission line tower was not considered during the pre-feasibility stage of project as they were far away from ROW. But as hill slope excavation started, hill slope started to slide, and civil agency representatives observed the cracks developed in foundation of towers as secondary/associated impact of landslide/ground failure. The civil agency had raised this issue with concern authorities. But during the winter rain, in year 2017, one transmission tower situated at Ch 71 + 300 km (zone of Bali landslide), 60 m far away from ROW, was suddenly collapsed due to ground sinking. Another adjustment tower was collapsed within one month

(33 m away from ROW). Due to collapse of these towers, some Kashmir valley districts were in dark for 4–5 days.

After the study, based on geological strata and slope failure prediction and physical condition of tower foundation, it was identified that 28 numbers of towers in critical condition in which 26 numbers were suggested to be protected by micro-piles, rock anchors, and cladding wall as per site condition, and two towers are suggested to shift at safer location. Some critical tower photographs are given below (Photographs 11–14):

Photograph 11 High-tension transmission tower near the ROW

Photograph 12 High-tension transmission tower near the ROW

Photograph 13
High-tension transmission tower near the ROW

Photograph 14
High-tension transmission tower near the ROW

5 Precautionary Measures for Landslide Zones

For developing road infrastructure in Himalayas, it needs hill excavation and hill slope, and ground failure cannot be ruled out. Based on type of failure observed in study area and geology selected, the mitigation measures construct the supporting structures (retaining/breast/cladding wall as required) and for old landslide zones such as Bali, Samroli, and Chenani areas, proper slope protection measures such as rock net, anchoring, shotcrete, rock fall barrier, bio-engineering methods with drainage hole for avoiding the pore water pressure are applied.

Photograph 15
High-tension transmission tower protected through cladding wall in study area

The safety of agriculture and residential ground to be taken care on high priority and measure through geotechnical instrumentation (crack meter, settlement markers, deep settlement markers, etc.) and protection of toe would be the high priority through slope protection structures. It is better to shift the probable residential houses at another safer place.

For protection of high-tension transmission tower near by the ROW, two type of protection measures adopted in the study area depending on the ground are available for protection. First is cladding wall (Photograph 15) with drainage hole and second one micro-piling with anchors, waler beam, and drainage holes. Micro-pile should be intact within rock and sufficient depth (prefer 5 m below the road level) and install the three side of tower except hill side. Micro-pile should be followed by shotcrete and drainage hole. Typical micro-piling section has been shown in Fig. 3 and Photographs 16 and 17.

Apart from above discussed mitigation measures, three stages of landslide mitigation measures need to be adopted during planning stage to operation stage of project, given below in Table 2.

5.1 Investigation Stage Mitigation Plan

During the study of project feasibility stage, following points need to be taken care:

- Old landslide and probable landslide zones are to be identified properly and avoid such zones for construction activities. If not feasible to avoid, natural slope protection measures are to be suggested as well as recommend bridge, tunnel as per site feasibility.
- A separate report needs to be prepared based on geological, geomorphological, geotechnical, hydrological factors that cover the impact of construction activity on landslide or natural slope throughout project section.

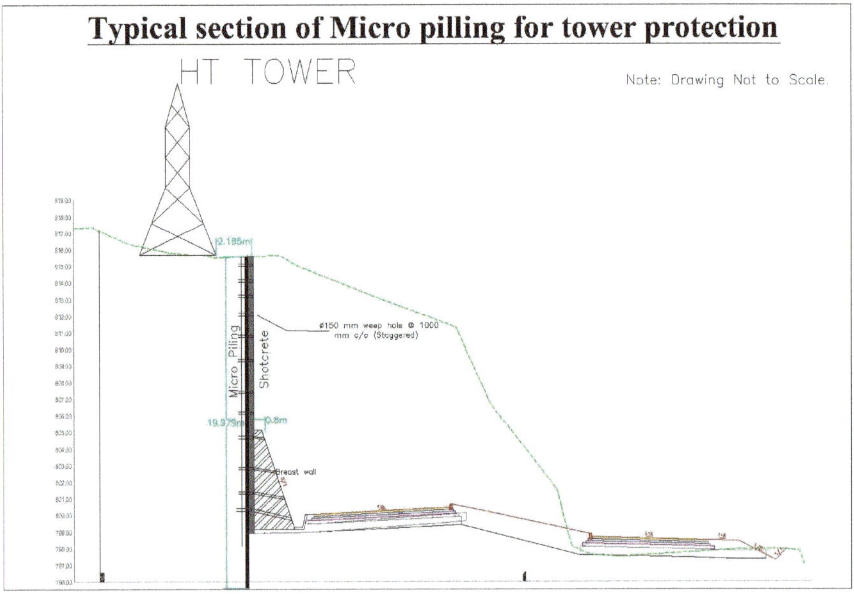

Fig. 3 Typical section of micro-piling for tower protection

Photograph 16 Tower protection with micro-pile and breast wall in study area

- Public awareness and government restriction to be physically marked after the landslide zonation mapping. No financial funding or other facilities are to be provided, if construction activities are done in that demarcated area.
- ROW for project is to be fixed as per the slope protection required land by authorities. Less land acquisition is cheaper during initial stage of project, but challenges faced during the construction and operational stage of project become much higher.

Photograph 17 Tower protection with micro-pile and breast wall in study area

5.2 *During Construction Stage*

- Stabilized the hill slope by cutting of slope with benches and retaining wall and Bio-engineering methods.
- Giving proper drainage for avoiding the ground failure. Catch water drain, drainage hole/weep holes, and diversion of water that are to be properly designed for free flow of pore water.
- All the site representatives must know the slope hazard zone and suggested mitigation/precaution measures.
- Project-approving agency must issue slope protection work as separate head to avoid incumbent financial and time load to construction contractor for smooth tackle of the slope protection work.
- As much as possible, avoid hazard-prone zone for disturbing the natural slope, and if disturbing is essential, then excavation is to be done with all precautionary measures.
- ROW if not adequate and slope heights are more, the mitigation measures are to be judiciously provided.

5.3 *During the Operation Stage*

- All slope protection works are to be done
- Instrumentation is to be fixed at critical zones for daily monitoring of ground stability and must issue warning at appropriate stages.
- Inputs from public awareness are always beneficial to identify the upcoming ground failure.

Table 2 Landslide protection during road construction

Landslide protection during road construction								
Investigation stage			Construction stage				Operational stage	
Identify and avoid the land slide zones	Hydrological study for avoid the damages due to flash flood	ROW should be sufficient for maintaining the natural slope	Scientific and quality method for excavating the slope and filling the muck	Prevention of landslide by bridges or tunnels	Slope protection measures to be applied immediately	Proper channelization of runoff from the upper slope	Monitoring of landslide prone area	Continue repairing/mitigation measures on landslide zone for smooth running of traffic

6 Conclusion

Widening of NH-44 is full of geo-environmental challenges. Climatic and hydrometeorological factors are playing as catalyst for landslide, hill slope and ground failure hazards during construction stage of project. Implementing precautionary measures step-by-step will reduce landslide risks. For NH-44 to work smoothly, various types of problems should be dealt with differently. Based on the types of landslides and related impacts in Udhampur–Chenani road section, it is observed that area is covered with thick debris materials with steep slope and water bodies (rainfall, nallas, and rivers). Natural features such as geology of area, geomorphology of steep disturbed slope, and hydrological factors of subsurface water percolation are predominantly present in the area of study. Combined effect of these natural features mainly causes landslide and related challenges during excavation.

Based on the direct and associated impact of ground failure in study area, it is suggested that development with modern scientific way can minimize the impact of failure. Independent probable ground failure study with mitigation measures is to be prepared during the pre-feasibility study of project. It should be content of stepwise slope protection measures, protection from wedge and planner failure, and drainage that would be appropriate for safety aspect from landslide hazard.

We have to develop hill development authority which prepares long-term plan for hill development in Himalaya after taking care of local geological challenges. Landslide study during the project feasibility stage is highly recommended with mitigation measures. Funds should also be allotted in infrastructure project for slope protection measures, and that would save the project completion time as well as project cost escalation. Economically, the protection measure cost may not be feasible at this movement, but in long term about maintenance, blockage of highway and impact of day to day life would be much cheaper.

References

Ahmad S, Alam A, Ahmad B, Bhat MI, Bhat MS (2015) Geomorphic evidence of nrecognized Balapur fault segment in the southwest Kashmir basin of northwest Himalayas. Geomorphology 250:159–172

Alam A, Ahmad S, Bhat MS, Ahmad B (2015a) Response to the comment by Shah, A. A. 2015 and further evidence supporting the dextral strike-slip pull-apart evolution of the Kashmir basin along the central Kashmir fault (CKF). Geomorphology 253:558–563

Alam A, Ahmad S, Bhat MS, Ahmad B (2015b) Tectonic evolution of Kashmir basin in northwest Himalayas. Geomorphology 239:114–126

Altaf S, Meraj G, Romshoo SA (2014) Morphometry and land cover based multi-criteria analysis for assessing the soil erosion susceptibility of the western Himalayan watershed. Environ Monit Assess 186(12):8391–8412

Ashraf Z, Nag RB (1966) A report on the detailed geological investigation of the Nashri Landslide on Jammu-Srinagar National Highway, JandK State. (Unpublished report). Geological Survey of India, Northern region, Lucknow

Bhat et al (2017) Geomorphological Field Guide Book on Kashmir Himalaya. In: 9th international conference on geomorphology of the International Association of Geomorphologists (IAG), New Delhi (6–11 November, 2017). http://geoinfo.amu.edu.pl/sgp/LA/LA35/B4.pdf
Bhat GM, Pandita SK, Dhar BL, Sahni AK, I Ul Haq (2002) Preliminary geotechnical investigation of slope failures along Jammu-Srinagar national highway between Batote and Banihal. Reprinted from Aspects of Geology Environment of the Himalaya, pp 275–288. http://www.scirp.org/(S(351jmbntvnsjt1aadkposzje))/reference/ReferencesPapers.aspx?ReferenceID=1417557
Chingkhei RK, Shiroyleima A, Robert Singh L, Kumar A (2013) Landslide hazard zonation in NH-1A in Kashmir Himayala, India. Int J Geosci 4:1501–1508
Meraj G et al (2015) Assessing the influence of watershed characteristics on the flood vulnerability of Jhelum basin in Kashmir Himalaya. Nat Hazards 77(1):153–175
Meraj G, Romshoo SA, Altaf S (2016) Inferring land surface processes from watershed characterization. In: Geostatistical and geospatial approaches for the characterization of natural resources in the environment. Springer, Cham, pp 741–744
Meraj G et al (2018) Geoinformatics based approach for estimating the sediment yield of the mountainous watersheds in Kashmir Himalaya, India. Geocarto Int 33(10):1114–1138
Nathawat MS et al (2010) Monitoring & analysis of wastelands and its dynamics using multiresolution and temporal satellite data in part of Indian state of Bihar. Int J Geomatics Geosci 1(3):297–307
Pandey Vinay Kumar (2018) Geological and geotechnical challenges In road widening, National Highway-44:-Jammu-Udhampur-Banihal-Qazigund, JandK, India. Int Res J Earth Sci 6(12):22–28
Rather MA et al (2017) Assessing the influence of watershed characteristics on soil erosion susceptibility of Jhelum basin in Kashmir Himalayas. Arab J Geosci 10(3):59
Shanker R et al (1989) Stratigraphy and sedimentation in Himalaya: a reappraisal. In: Geology and tectonics of Himalaya, G.S.I, Spl Pub 26:1–60
Singh Y (2006) Geotechnical and structural evaluation of tectonostratigraphic units along the National Highway, NH-1A between Udhampur and Batote Jammu Himalaya. (Unpublished doctoral dissertation), University of Jammu, Jammu, India
Thakur VC, Jayangondaperumal R, Malik MA (2010) Redefining Medlicott–Wadia's main boundary fault from Jhelum to Yamuna: an active fault strand of the main boundary thrust in northwest Himalaya. Tectonophysics 489:29–42
Verma RS (1966–67) Geological report on the landslides along the Jammu-Srinagar national highway (1A), Jammu and Kashmir State. Geological Survey of India, Northern region, Lucknow. Field Season report 1966–67. http://www.portal.gsi.gov.in/gsiDoc/pub/NRO-JK_12925.pdf

Floods in the Himalayas—Causes and Consequences

September 2014 Floods in Kashmir Himalaya—Impacts and Mitigation Strategy

Mohammd Rafiq, Amit Parashuram Kesarkar, Usamah Derwaish, and Aaqib Manzoor Bhat

Abstract Floods are the most damaging disasters in the world. The causes of the floods may be natural or anthropogenic, but people have to bear the burden of devastation in both cases. Most cities and urban areas are located in or near flood plains due to their suitability for agriculture and urban development due to the flat and fertile nature of the soil. However, to make use of these facilities, people have to live in these flood-prone areas. The forecast of the water levels of a flood-prone river must be made to minimize damage to life and property. Prediction and early warning systems can also help the concerned authorities to develop mitigation strategies to reduce the impacts of floods. In this book chapter, one such flood event is discussed, which occurred in the union territory of Jammu and Kashmir in September 2014. We have analyzed the reasons and the effect of this flood event, and the mitigation strategies are also discussed. Extreme rainfall, which exceeded by more than 1200% from the normal rainfall at various districts of Kashmir valley during the first week of September 300%, decrease in the snow-covered area in less than ten days over Kashmir valley was observed. This snowmelt contributed to the cumulative discharge over the region. We suggested long-term mitigation measures by developing an efficient early warning system and enhancing the watershed management practices using state-of-the-art remote sensing and GIS techniques.

Keyword Kashmir 2014 floods. Early warning system. Flood mitigation

---

M. Rafiq (✉) · A. P. Kesarkar
Department of Space, National Atmospheric Research Laboratory, Govt. of India, Gadanki, Andhra Pradesh 517112, India

U. Derwaish
Department of Civil and Environmental Engineering, Politecnico di Milano, 20133 Milan, Italy

A. M. Bhat
Department of Rural Development and Panchayati Raj, Government of Jammu and Kashmir, Jammu and Kashmir 190001, India

S. Kanga et al. (eds.), *Disaster Management in the Complex Himalayan Terrains*, Geography of the Physical Environment,
https://doi.org/10.1007/978-3-030-89308-8_6

1 Introduction

Due to the complex orography of Kashmir valley, floods, snow avalanches, landslides, etc. are common and have a long history of floods. The low-level convergence of south easterlies and northwesterly coupled with orographic uplift causes very heavy rainfall, making Kashmir a flood-prone region. Kashmir valley has a long history of floods, and a flood of 1983 (Lawrence 1895) was one of the major floods documented. This flood had severe impact on agriculture as it divested the rice plantation that was ready to be harvested. This flood is known for food crises, and 25,426 acres of crops were submerged also as reported by Lawrence 1895. Many people died due to hunger; he has also estimated that almost half of the population of Jammu and Kashmir was affected by this devastating flood (Khalid 2016). The rainfall occurred continuously for 52 h., which resulted in this great flood of 1893. In 1903, another major flood was recorded in the valley, which destroyed 7000 houses (Mishra 2015). Some of the major floods occurred in 1929, 1948, 1950, 1957, 1959, 1992, 2010, and 2014, which will be further discussed in this chapter. A case study of the 2014 flood, dealing with understanding the causes, impact, and future strategies for being prepared for any possible future similar event, will be discussed further in this chapter.

2 Causes and Impact

The devastating flood of September 2014 was triggered by the extreme rainfall, which lasted for almost a week. Such an extreme flooding was not witnessed in the valley during past 60 years. This unprecedented flood claimed over 300 lives and destroyed properties worth millions. The losses are estimated to be more than $US6718.583 million. Almost, 5642 villages and 125,000 families have been affected. The capital city (Srinagar) suffered for the most part as it remained submerged in almost 18 feet of water for more than three weeks; people were trapped in their houses. More than 350,000 structures, mostly residential houses, have been damaged. (Tabish and Nabil 2015). Satellite data analysis shows that the area underwater/inundated was almost 557 km^2; 12% of this inundated area contained built-up, and 80% was the agricultural area (Bhatt et al. 2017a; b). The built-up land over Dal Lake (a low-lying lake in the capital city of Jammu and Kashmir) was decreased to 20.25^2 km from 25.44 km^2 (Ahmad et al. 2020).

Rainfall observations showed a strong diurnal variation from September 2–8, 2014. Analysis of the synoptic conditions contributing to unprecedented rainfall reveals that the rains were triggered by the interaction of the westward-moving low-pressure monsoon through central and northwestern India and the eastward-moving deep trough in the mid-tropospheric westerlies. The additional low-pressure areas developed over Saurashtra and Kutch on September 3, 2014 and over the Bay of Bengal on September 5, 2014 ensured that the event was sustained by heavy wind and moisture flow in Jammu Kashmir (Ray et al. 2015). Satellite data revealed that the rainfall rates in the range of 15–22 mm/h occurred during the September 2–8, 2014. The results show that the cumulative rainfall during September 2–6, 2014, may have contributed to the flood events (Mishra 2015; Ray et al. 2015).

Bhatt et al. (2017a; b) explored the cause of the September flood by highlighting the effect of rainfall and pointed out the lack of a flood channel to drain the water rapidly quickly. The flat topography of the valley does not allow the accelerated overflow of the floodwaters; hence, the capital city (Srinagar) and other northern parts remained inundated for more than two weeks (Kumar & Acharya 2016). An integrated solution, taking into account the hydrological, geological, and geomorphological dimensions of the river basin, is essential for the mitigation policies. Meraj et al. (2015a, b) evaluated the effect of land cover, slope, and morphometry, on flood risk in downstream areas of the two watersheds of the Jhelum basin and highlighted their important role (Meraj et al. 2016). A strong influence on the hydrological response and functionality is governed by the geomorphology, topography, and land use. He suggested exploring all the watersheds of the Jhelum River for hydrological responses using remote sensing and GIS aided by the field data, which can help in long-term mitigation of floods. Romshoo et al. (2018), highlighted the impact of anthropogenic drivers on the September 2014 floods. The widespread urbanization of floodplains, the depletion of wetlands, and the decreased drainage capacity of Jhelum due to silting played an important role in the floods which were triggered by extreme rainfall. He pointed out an inadequate flood management system and a lack of structural capability to handle the severity of significant flood incidents.

However, there was no such study which has measured the contribution of snowmelt to these floods. In 2013–2014 winter, Kashmir valley received a good amount of snowfall. Many studies have reported extreme changes in the snowfall patterns and extreme snowfall events over this region (Mishra and Rafiq 2017a; Rafiq and Mishra 2018a, Rafiq et al. 2021). There was a decrease of 300% in the snow-covered area (SCA) from August 29 to September 7, 2014, which mainly contributed to the floods. Satellite data revealed a decrease of 1590 in SCA in less than a week. The decrease in snow-covered area was witnessed in the Kedarnath floods 2013; also, there was a decrease of 50% in the SCA within a week which contributed to the filling of a glacier lake (Chorabari Lake), and hence, it breached catastrophically producing a discharge of 1700 $m^3\ s^{-1}$. Uttarakhand flood 2013 was also triggered by extreme rainfall events and aided by the rapid snowmelt (Rafiq et al. 2019b).

Unfortunately, till date, no such study has been carried out in the valley of Kashmir describing the equivalent amount of liquid water stored in the snowpack. Snow Water Equivalent (SWE) data can theoretically show the amount of water discharged from the snowmelt, which can be useful to build an early warning system. This data, along with the precipitation forecast, can help to build an efficient flood warning system. This will also require an efficient discharge measuring station network and a well-established rain gage data. Kashmir valley has a very poor rain gage and discharge measuring network. There are only six IMD station measuring various atmospheric parameters including precipitation and temperature. It is noteworthy that the rain gage distribution density over most flood-affected districts is so poor that there are 0–2 rain gages in a 200 km^2 area Table 1, and many of the districts do not have any

Table 1 List of the rain gages over Kashmir (2012 updated)

Districts	Number of stations
Budgam	4
Baramulla	10
Anantnag	11
Srinagar	5
Pulwama	3

station. This poor density of rain gage stations and the complex topography limit the applicability of rain gages to monitor rainfall during flood events in Kashmir (Mishra 2013). Moreover, a few of these rain gage stations stopped working during heavy rain events. The same is the case with discharge station three are only 3 discharge measuring stations along the river Jhelum which has a catchment area of about 16,000 km^2.

In order to mitigate such events in the future, we need a well-established early warning system based not only on the precipitation estimates but also the data about morphometry, land use, topography, temperature, SWE, etc.

3 Study Area

We have studied the September 2014 flood which occurred in Kashmir valley. The figure below (Fig. 1) displays the study area; also, the 24 watershed of river Jhelum flowing through the union territory of Jammu and Kashmir have been shown in the figure (Fig. 1). The river flows through the heart of Kashmir valley and acts as lifeline for irrigation and drinking water for a huge population. The river originates from a spring known as Verinag (Coward et al. 1972). The Jhelum River is also fed by several tributaries along its course from Pir-Panjal on the left and from the Himalayan range on the right. The drainage network along with the Himalayan and Pir-Panjal mountain range on both sides of the valley has also been shown in Fig. 1.

Vishav watershed has the highest water-yielding potential, and also, it is the fastest water-yielding catchment of the Jhelum basin. It is followed by Bringi, Lidder, Kuthar, Sind, Madhumati, Rembiara, Sukhnag, Dal, Wular-II, Romshi, Sandran, Ferozpur, Viji-Dhakil, Ningal, Lower Jhelum, Pohru, Arin, Doodganga, Arapal, Anchar, Wular-I, Gundar, and Garzan in the case of a same intensity storm event (Meraj et al. 2018a, b, c).

4 Results and Discussion

Multi-satellite sensors, observational, and model data are used in this study to access the contribution of cryospheric (mostly snow/glacier) changes on land surface

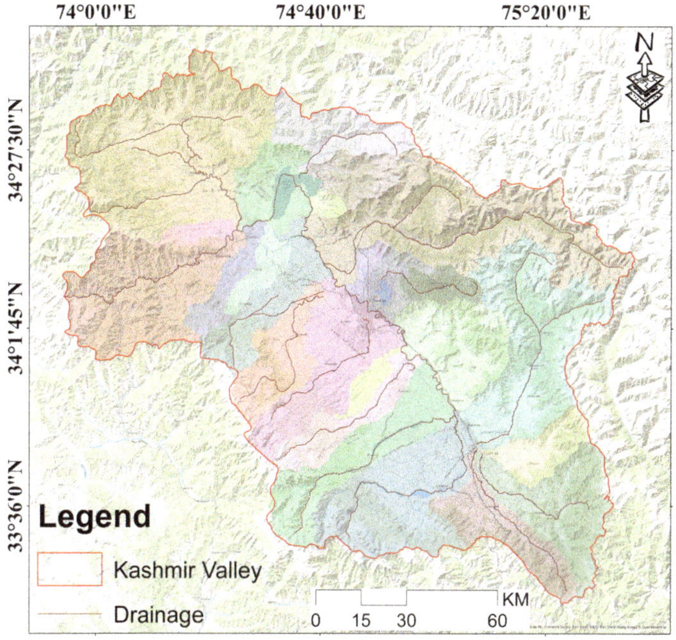

Fig. 1 Displays the study area and watersheds of the Jhelum River basin

processes, including the hydrology (Nathawat et al. 2010; Altaf et al. 2013; Pall et al. 2019; Romshoo et al. 2020). The Landsat data (Table 2) are used to derive the snow and glacier changes over the study area, and the ancillary data are also used to justify the results. Also, the MODIS MOD10A2 (Table 3) is used to analyze SCA. Daily as well as 8-day composite images from MODIS are used to derive the SCA and validated with Landsat data. There are a number of studies on the floods which occurred in the union territory of Jammu and Kashmir as stated in the introductory part above. However, no such study has been carried out in the study area which has accessed the contribution from the snowmelt to these devastating floods. Rafiq et al. (2019b) have mentioned the role of snowmelt in the Kedarnath floods of June 2013 where almost 50% snow cover declined in 4–5 days over the Kedarnath area. We have analyzed the SCA changes during the September flood of 2014 in the union territory of Jammu and Kashmir, and we found that more than 1590 km^2 of snow-covered area melted in less than a week. This snowmelt has contributed to the floods which was triggered by the extreme rainfall events. Heavy rainfall rates of 15–22 mm/h occurred over the UT of Jammu and Kashmir during the first week of September 2014. About 24 mm/h rainfall occurred over the Anantnag and the Kulgam districts of the Kashmir valley. However, the maximum rainfall (hourly) never exceeds the 25 mm/h threshold, thus excluding the possibility of cloud burst. Rainfall in the range 400–500 mm was accumulated over the districts of Pulwama and Anantnag. These extreme rainfall events may have accelerated the snowmelt as in the winter season of 2013–2014 Jammu and Kashmir received a good amount of snowfall.

The area under snow cover on August 29, 2014 was 2379.6 km^2 which decreased to 789.3 km^2 on September 7, 2014 (Fig. 2). Landsat data also displayed the same trend (Fig. 3). The areas with maximum change are highlighted with white eclipse. In these areas, almost all the snow melted during this flood event. The rainfall played an

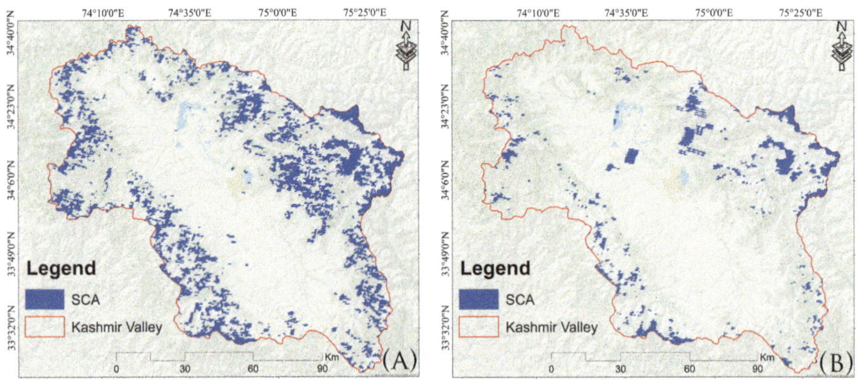

Fig. 2 MODIS snow-covered area (SCA) over Kashmir valley on 07–29-2014 (a), and 08–07-2014 (**b**)

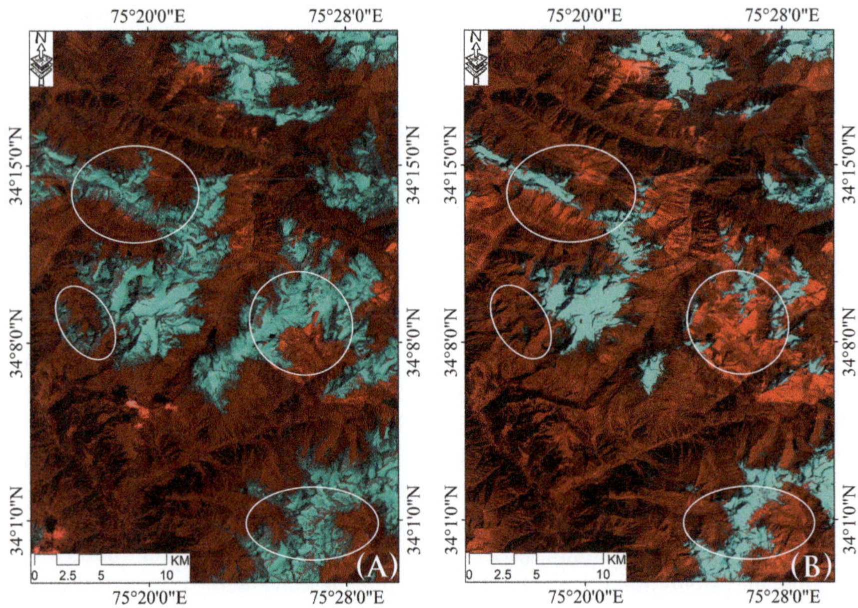

Fig. 3 Snow cover from a false composite of Landsat on 08–25-2014 (**a**) 09–19-2014 (**b**)

important role in accelerating the snowmelt. There was almost a 300% decrease in the SCA within a week. This large amount of snowmelt enhanced the discharge, which was already increased due to extreme rainfall. The rainfall data during the first week of September 2014 over the union territory of Jammu and Kashmir showed a strong diurnal variation. The rainfall rate was almost 20 mm h^{-1} in most of the districts. Kulgam and Anantnag received almost 200 and 150 mm of rainfall on September 5, 2014. The accumulated rainfall during this week was above 200 mm in most of the districts. Kulgam received almost 600 mm of rainfall, which was followed by Anantnag receiving almost 500 mm rainfall (Rashid et al. 2019). Shopain received about 300 mm, and the capital city Srinagar recorded almost 170 mm of rainfall (accumulated) during the first week (1–7) of September 2014. Anantnag received 1222% above the normal rainfall. The normal range for September at Anantnag is 32.9 mm, and it received 402.3 mm.

Similarly, Kulgam received 540.5 mm of rainfall exceeding 1259% from the normal rainfall of 42.9 mm. Shopain and Pulwama also received the above normal rainfall exceeding by 1390% and 1470%, respectively. During September for Shopain and Pulwama, the normal rainfall is 29.2 mm and 19.9 mm; they received 406 mm and 292.7 mm, respectively.

The discharge station located at Sangam recorded the highest flood level of 1595, and 1589 m was recorded at Srinagar (Ram Munshibagh), which is almost 11 m and 9 m above the reference point (mean sea level of zero), respectively. Nearly, 3.5% of the Kashmir valleys geographical area was under flood. Srinagar district was severely affected and remained inundated for almost three weeks. About 287 villages with 22 lakh people were affected. About 444 km^2 of agricultural land was affected, and 20 km^2 of horticulture land and 67 $km^2$of built-up area were affected (Bhatt et al. 2017a; b).

5 Conclusion, Strategies, and the Way Forward

Floods are one of the most havoc creating and devastating disasters (Joy et al. 2019). The loss of life and property every year signifies the need for an early warning system that could warn the concerned community beforehand to relocate to safe areas. As the future projection show an increase in the frequency of extreme precipitation events (Gujree et al. 2017a, b Rafiq et al. 2019a, Mishra et al. 2019, Mishra and Rafiq 2019), there is a need to have recovery planning related to the long-term stability, which keeps displaced residents from collapsing into homelessness and to provide them with resources for economic growth through better reconstruction (Agarwal et al. 2014). There are several early floods warning systems developed which use different technologies and methods to suit different scenarios (Kaushik et al. 2020). All those models require a sufficient amount of dynamical observational data to predict the floods accurately, which include accurate temperature and precipitation forecasts (both solid and liquid), changes in the snowmelt (SCA) (Martinec 1986, Kanga et al. 2020). The data like slope, land use, and other morphometric data such as lag time are also required to train the model at its initial phase. There is a need to improve weather predictions in the highly complex orographic regions like Kashmir (Mishra and Rafiq 2017b,2018b) and develop an effective warning system. For that, there is a need in the union territory of Jammu and Kashmir for expansion of the rain gage stations as well as the discharge measuring stations. For long-term flood control, an integrated solution is required, taking into account the hydrological, geological, and geomorphological dimensions of the river basin. Depending on the characteristics of the watershed, the hydrological reaction has a considerable effect on the frequency and severity of the intensity of the floods and needs to be better understood (Meraj et al. 2018a, b, ;c Rather et al. 2017). Non-structural steps, including watershed maintenance activities, need to be placed in place to absorb excess runoff. The valley has several natural wetlands, but they are quickly declining, which need to be restored to contain excess floodwaters and serve as a "cushion" to preserve the region. In order to mitigate the effects of such disasters in the future, there is a clear need to improve flood early warning services assisted by well-distributed hydro-meteorological networks (including Doppler weather radars) and river discharge stations.

Other measures for long-term mitigation of floods such as 2014 in the valley of Kashmir require the following: -

- Increasing the carrying capacity of river Jhelum to 35,000 cusecs, and further, 15,000 cusecs are added through the flood channels. The 2014 flood of September produced a discharge of 1,35,000 cusecs. There is a need to create flood storage dams on the tributaries of the Jhelum River, which can store the access water during floods.
- Each watershed of Jhelum has a unique identity and needs different planning, which can be analyzed using the remote sensing and GIS technique. The planning of watersheds should be carried out to view its geomorphology, topography, and land use, which can help in long-term mitigation. River management practices like bank protection and strengthening, dredging, and planning new embankments, and their proper monitoring can help in the long-term mitigation process.
- Assessment of flood hazard maps using remote sensing and GIS approaches is required and needs to be placed in the planning sector. It can be achieved using the extracted information regarding the flood area extent and depth from previous floods as a reference. The modeling techniques supplied with a good number of observational data sets can help in simulating and demarking the areas, especially the low-lying areas, which act as a detention basin during floods.
- The prediction of water levels of a flood-prone river is must in order to abate the life and property damages. The losses happen every year which indicates the need of an early warning system, which could inform the concerned community early so that they could move to the safe areas. Genetic algorithms can be used to train the models which are helpful in predicting the discharge (Imran et al. 2021).

There should be a proper recovery management plan which can help in post-flood recoveries. This should be governed by a competent authority that can facilitate sequenced and prioritized direction for rehabilitation and reconstruction. The post-development should meet the merits of sustainable development so that the related circumstances are not caused in the future. This will require an enhanced flood-forecasting system, updating and creation of building regulations, flood zonation, updating of land use land cover maps, and also the disaster resilient construction. The panel created for this recovery management should include the experts from various sectors which can assist in the various phases of recovery and rehabilitation, e.g., engineers, financial and GIS experts, meteorologists, hydrologists. Finally, an appropriate financial and institutional support with freedom of collaboration should be developed for its overall implementation.

Acknowledgements We are thankful to the Director, National Atmospheric Research Laboratory, India, for providing the laboratory facilities and fellowship offered to the corresponding author. We also acknowledge the Srinagar IMD and Flood and Irrigation department Jammu and Kashmir for providing the observational data. The MODIS SCA daily L3 and 8-day composite MOD10A2 datasets were acquired from the Atmosphere Archive & Distribution System (LAADS) Distributed Active Archive Center (DAAC), located in the Goddard Space Flight Center in Greenbelt, Maryland (https://ladsweb.nascom.nasa.gov/).

References

Agarwal S, Fulzele TU, Aggarwal G (2014) Flood recovery management in Jammu and Kashmir: a tool for resilience. Asian J Environ Disaster Manag 6(3):215–229

Ahmad T, Pandey AC, Kumar A (2020) Impact of 2014 Kashmir flood on land use/land cover transformation in Dal Lake and its surroundings, Kashmir valley. SN Appl Sci 2(4):1–3

Altaf F, Meraj G, Romshoo SA (2013) Morphometric analysis to infer hydrological behaviour of Lidder watershed, Western Himalaya, India. Geogr J 2013

Bhatt CM, Rao GS, Farooq M, Manjusree P, Shukla A, Sharma SV, Kulkarni SS, Begum A, Bhanumurthy V, Diwakar PG, Dadhwal VK (2017) Satellite-based assessment of the catastrophic Jhelum floods of September 2014, Jammu & Kashmir, India. Geomatics Nat Hazards Risk 8(2):309–27

Bhatt CM et al (2017) Satellite-based assessment of the catastrophic Jhelum floods of September 2014, Jammu & Kashmir, India. Geomat Nat Haz Risk 8(2):309–327

Coward JM, Waltham AC, Bowser RJ (1972) Karst springs in the Vale of Kashmir. J Hydrol 16(3):213–23

Gujree I et al (2017) Evaluating the variability and trends in extreme climate events in the Kashmir Valley using PRECIS RCM simulations. Model Earth Syst Environ 3(4):1647–1662

Gujree I et al (2017b) Evaluating the variability and trends in extreme climate events in the Kashmir Valley using PRECIS RCM simulations. Model Earth Syst Environ 3(4):1647–1662

Imran M, Khader PSA, Rafiq M, Rawat K S (2021). Forecasting water level of Glacial fed perennial river using a genetically optimized hybrid machine learning model. Mater Today Proc https://doi.org/10.1016/j.matpr.2021.02.256

Joy J, Kanga S, Singh SK (2019) Kerala flood 2018: flood mapping by participatory GIS approach. Meloor Panchayat. Int J Emerging Techn 10(1):197–205

Kanga S et al (2020) Modeling the spatial pattern of sediment flow in lower Hugli Estuary, West Bengal, India by Quantifying Suspended Sediment Concentration (SSC) and Depth Conditions using Geoinformatics. Appl Comput Geosci 100043

Kaushik S, Rafiq M, Joshi PK, Singh T (2020) Examining the glacial lake dynamics in a warming climate and GLOF modelling in parts of Chandra basin, Himachal Pradesh, India. Sci Total Environ 14:136455

Khalid W (2016) Media propaganda and the Kashmir dispute: a case study of the Kashmir floods. Reuters Institute for the Study of Journalism. University of Oxford, UK

Kumar R, Acharya P (2016) Flood hazard and risk assessment of 2014 floods in Kashmir Valley: a space-based multisensor approach. Nat Hazards 84(1):437–464

Martinec J, Rango A (1986) Parameter values for snowmelt runoff modelling. J Hydrol 84(3–4):197–219

Meraj G, Romshoo SA, Altaf S (2016) Inferring land surface processes from watershed characterization. In: Geostatistical and geospatial approaches for the characterization of natural resources in the environment. Springer, Cham, pp 741–744

Meraj G et al (2018b) An integrated geoinformatics and hydrological modelling-based approach for effective flood management in the Jhelum Basin, NW Himalaya. Multi Digital Publis Inst Proc 7(1):8

Meraj G, Khan T, Romshoo SA, Farooq M, Rohitashw K, Sheikh BA (2018a) An integrated geoinformatics and hydrological modelling-based approach for effective flood management in the Jhelum Basin, NW Himalaya. Multi Digital Publis Inst Proc 7(1):8

Meraj G, Romshoo SA, Yousuf AR, Altaf S, Altaf F (2015a) Assessing the influence of watershed characteristics on the flood vulnerability of Jhelum basin in Kashmir Himalaya. Nat Hazards 77(1):153–175

Meraj G et al (2015b) Assessing the influence of watershed characteristics on the flood vulnerability of Jhelum basin in Kashmir Himalaya. Nat Hazards 77(1):153–175

Meraj G et al (2018c) Geoinformatics based approach for estimating the sediment yield of the mountainous watersheds in Kashmir Himalaya, India. Geocarto Int 33(10):1114–1138

Mishra A, Rafiq M (2017) Towards combining GPM and MFG observations to monitor near real time heavy precipitation at fine scale over India and nearby oceanic regions. Dyn Atmos Oceans 1(80):62–74

Mishra AK (2015) A study on the occurrence of flood events over Jammu and Kashmir during September 2014 using satellite remote sensing. Nat Hazards 78(2):1463–1467

Mishra AK, Nagaraju V, Rafiq M, Chandra S (2019) Evidence of links between regional climate change and precipitation extremes over India. Weather 74(6):218–221

Mishra AK, Rafiq M (2017) Analyzing snowfall variability over two locations in Kashmir, India in the context of warming climate. Dyn Atmos Oceans 1(79):1–9

Mishra AK, Rafiq M (2019) Rainfall estimation techniques over India and adjoining oceanic regions. Curr Sci 10(10):56–68

Nathawat MS et al (2010) Monitoring & analysis of wastelands and its dynamics using multiresolution and temporal satellite data in part of Indian state of Bihar International J Geomatics Geosci 1(3):297–307

Pall IA, Meraj G, Romshoo SA (2019) Applying integrated remote sensing and field-based approach to map glacial landform features of the Machoi Glacier valley, NW Himalaya. SN Appl Sci 1(5):488

Rafiq M, Mishra AK (2018) A study of heavy snowfall in Kashmir, India in January 2017. Weather 73(1):15–17

Rafiq M, Mishra AK (2018) Remote sensing of near-real-time heavy precipitation using observations from GPM and MFG over India and nearby oceanic regions. Multid Digit Publishing Inst Proc 2(7):327

Rafiq M, Mishra AK, Panda J, Sharma SK (2019a) monitoring convective clouds over india and nearby regions using multi-spectral satellite observations. In: Proceedings of international conference on remote sensing for disaster management. Springer, Cham, pp 51–60

Rafiq M, Para JA, Kesarkar A, Lotus S (2021) A study of extreme snowfall during 2019 and 2020 across the Kashmir Himalaya. Weather 76. https://doi.org/10.1002/wea.3938

Rafiq M, Romshoo SA, Mishra AK, Jalal F (2019b) Modelling Chorabari Lake outburst flood, Kedarnath, India. J Mountain Sci 16(1):64–76

Rashid I, Parray AA, Romshoo SA (2019) Evaluating the performance of remotely sensed precipitation estimates against in-situ observations during the september 2014 mega-flood in the Kashmir Valley. Asia-Pac J Atmos Sci 55(2):209–219

Rather MA et al (2017) Assessing the influence of watershed characteristics on soil erosion susceptibility of Jhelum basin in Kashmir Himalayas. Arab J Geosci 10(3):59

Ray K, Bhan SC, Bandopadhyay BK (2015) The catastrophe over Jammu and Kashmir in September 2014: a meteorological observational analysis. Curr Sci 10:580–591

Romshoo SA, Altaf S, Rashid I, Dar RA (2018) Climatic, geomorphic and anthropogenic drivers of the 2014 extreme flooding in the Jhelum basin of Kashmir, India. Geomat Nat Haz Risk 9(1):224–248

Romshoo SA et al (2020) Satellite-observed glacier recession in the Kashmir Himalaya, India, from 1980 to 2018. Environ Monit Assess 192(9):1–17

Tabish SA, Nabil S (2015) Epic tragedy: Jammu & Kashmir floods: a clarion call. Emerg Med (los Angel) 5(233):2

A Review on the Estimation of Glacial Lake Outburst Floods (GLOFs) in the Himalayan Region Using Remote Sensing and Geographic Information System

Mohd Talha Anees, Naseem Akhtar, Ahmad Farid Bin Abu Bakar, and Muhammad Izzuddin Syakir Ishak

Abstract Glacier lake outburst flood (GLOF) is one of the most devastating natural hazards in the Himalayan region. Several studies used identification and size monitoring of glacier lakes in the Himalayan region using remote sensing and GIS. For glacier lake bathymetry, most of the studies used in situ data. However, the collection of bathymetry data of several glacier lakes in the Himalayan region is difficult, time-consuming, and expensive. None of the studies used remote sensing data to estimate the bathymetry of glacier lakes. In this chapter, different types of floods in the Himalayas and the use of remote sensing data to estimate river/coastal bathymetry were reviewed. Furthermore, the application of remote sensing data to estimate glacial lake bathymetry is also discussed.

Keywords Remote sensing · GIS · Glacier lake · Glacier lake outburst floods · Bathymetry

1 Introduction

Flood is one of the most common natural hazards in the Himalayan region and causes loss of lives in downstream valleys, economic losses, damages to infrastructures, intense land use changes, and drinking water contamination. It occurs mostly due to extreme rainfall events or climate change. The climate of the Himalayan region varies with elevation. Low-altitude region such as foothills of the Himalayas (60 m) gets tropical climate, while high-altitude region such as Mt. Everest (8848 m) which is 90–120 km to north gets colder climate (Nandargi and Dhar 2011). Wind systems

M. T. Anees (✉) · A. F. B. A. Bakar
Department of Geology, Faculty of Science, University of Malaya, 50603 Kuala Lumpur, Malaysia

N. Akhtar · M. I. S. Ishak
School of Industrial Technology, Universiti Sains Malaysia, 11800 Minden, Pulau Penang, Malaysia

S. Kanga et al. (eds.), *Disaster Management in the Complex Himalayan Terrains*, Geography of the Physical Environment,
https://doi.org/10.1007/978-3-030-89308-8_7

such as snowstorms, a mesoscale cyclonic storm, and other high-speed winds also strongly influence climate in the Himalayas, resulting in flash floods in this region (Nandargi and Dhar 2011).

Rainfall events are associated with variable altitude in the Himalayas. Maximum rainfall is received at the foothills of the Himalayas, second maximum is in the middle Himalayas (2400 m a.s.l), and then it decreases sharply toward higher elevations till the Great Himalayan Range reached (Nandargi and Dhar 2011). Literature showed an unexpected declining trend of rainfall events while increasing the trend of extreme rainfall and temperature. The increasing trend significantly enhanced the glacial melting in the Himalayas which is contributing huge discharge in various major river's headwater and, hence, flood in their downstream (Romshoo et al. 2020; Kumar et al. 2020). Black carbon is also a key factor in increasing temperature trends that influence the climate of the Himalayas. Black carbon is also contributing to glacial melts in the Himalayas.

Factors influencing the climate of the Himalayas are reviewed by Das and Meher (2019). The focus of this review is on the flash floods in the Himalayas, especially glacier lake outburst floods (GLOFs), their implications, assessing methods, and estimation of glacier lake bathymetry using an integrated approach of remote sensing and GIS. Therefore, the objectives of this study are (1) to review causes of GLOFs, their implications, and assessment methods, (2) to review methodologies of estimating bathymetry or unknown submerged topographies using remote sensing and GIS, and (3) discussing mitigation measures for averting the potential hazard associated with GLOFs using cases.

2 Methodology

For the first objective, the causes of GLOFs and their assessing methods were discussed. For the second objective, studies that used remote sensing data to estimate river bathymetry were reviewed. Remote sensing data include standard photogrammetry, multispectral imageries, digital elevation model (DEM), LIDAR-based DEM, passive optical imagery, and multiple band ratio. For the fourth objective, a summary of the implication of remote sensing application in glacial lake bathymetry estimation is mentioned (Fig. 1).

2.1 *Glacier Lake Outburst Floods (GLOFs)*

Extensive glacier shrinkage due to climate change is causing the formation of glacier lakes in low elevated areas of the Himalayan region. Due to high precipitation in the Himalayan region, sometimes, these lakes overflow or outburst due to failure of moraines-dammed. These types of flash floods in the Himalayan region are known as glacier lake outburst floods (GLOFs). GLOF is one of the most devastating natural

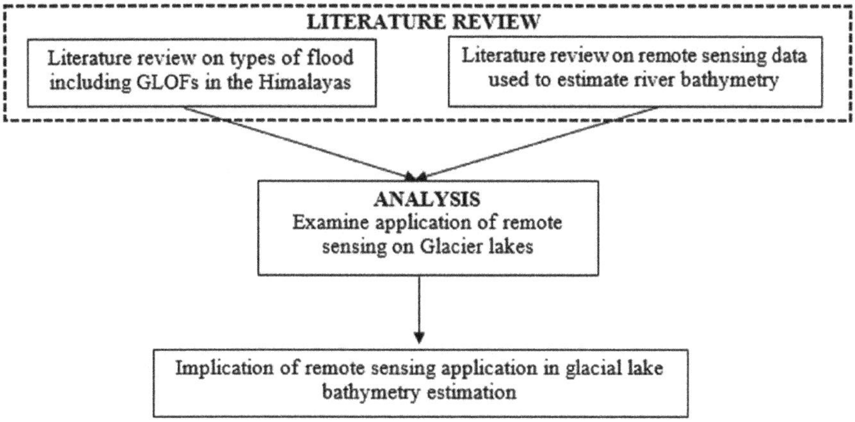

Fig. 1 Methodology workflow

hazards in the Himalayan region. It causes disastrous consequences by releasing large volumes of water in a very-short interval of time due to dam overflow and moraine or ice dam failure.

Different types of flash floods in the Himalayan region showed that these are directly related to the hydro-meteorological condition of the region which triggers other events such as landslides, debris flow, avalanches. The rise of temperature due to climate change plays an important role in glacial lake outbursts, sudden glacial and ice melting which causes flash flood at the downstream valley.

3 Glacier Lake Outburst Floods (GLOFs) in the Himalayas

Several GLOFs were detected in the Himalayan region in the past, which caused catastrophic destruction and fatalities in downstream regions. Some major GLOFs examples are in the Indrawati valley (the year 1956) (Muller 1959), Punakha flood, Bhutan (the year 1957), Dig Tsho below Langmoche glacier in Khumbu, Nepal (the year 1984) (Fushimi et al. 1985; Galay 1985, Ives 1986), Lugge Tsho, Nepal (the year 1994) (Watanabe and Rothacher 1996), Punakha Wangdue valley (the year 2000), Sabai Tsho in 1998 (Osti and Egashira 2009), and Chorabari in 2013 (Allen et al. 2016; Kanti et al. 2019). Furthermore, the phenomenon of GLOFs has been recorded as a recent event in the 2020s decades in the Indian Himalayas in Kedarnath. This event has caused the existence of thousands of human beings and a great loss of infrastructure and property in downstream areas such as Gobindghat, Rudraprayag City Rambara, and many more areas of the Rudraprayag District.

Recent studies identified and mapped active glacier lakes which are very important for GLOFs vulnerability and risk assessment. Allen et al. (2016) study the contributing factors of the Kedarnath disaster, Uttrakhand, in the year 2013. They

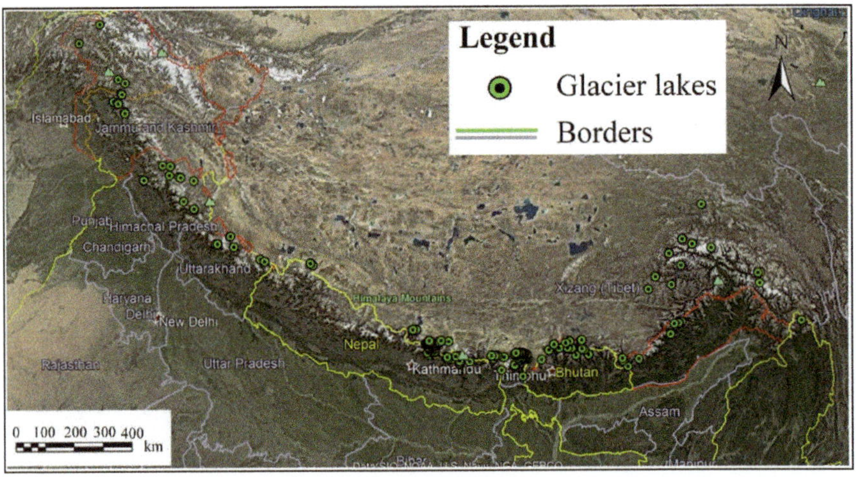

Fig. 2 Google earth image showing some active glacier lakes based on the literature

mentioned ten lakes with the most unfavorable topographic disposition in the area. Veh et al. (2018) detect glacier lakes using Landsat data and mentioned 11 documented and 10 detected lakes spread over the Hindu Kush Himalayan region. Dubey and Goyal (2020) identified 329 lakes in Indian Himalayas for a remote sensing-based hazard and risk assessment of GLOFs. Veh et al. (2019) assess GLOFs in the Hindu Kush–Karakoram–Himalaya–Nyainqentanglha region and mentioned 62 glacier lakes. Liu et al. (2020a, b) assess GLOF hazard in the Bhote Koshi Basin which is affected by the Gorkha earthquake. They mentioned 103 very high and high hazard glacial lakes. Some of the active glacier lakes mentioned in these studies are shown in Fig. 2.

3.1 Data Availability for GLOFs Assessment

Studies related to GLOF events are of three types such as overflow dam, moraine, and ice dam failure. There are two most important controlling factors of GLOF events such as the melting of the glacier due to climate change and high precipitation. Both factors are responsible for raising the water level in glaciers lake. Accurate prediction of GLOF could be possible through physically based models with precise input data. The collection of precise input data is affected due to difficulty in using logistics in remote areas and low-density observed data. To date, satellite remote sensing data has been widely used for the estimation of spatio-temporal changes in glacier lakes (Sharma et al. 2018; Dubey and Goyal 2020). Previous studies used Landsat images, IRS LISS III and IV, ASTER GDEM, and SRTM DEM. Landsat and IRS images were used to map glacier lakes while DEMs for slope delineation, morphometric parameter estimation, and topographic parameter estimation. All studies used in situ bathymetry data of glacier lake to measure the volume and area of the lake. Temporal in situ bathymetry data collection would be very expensive, laborious, and time-consuming.

Das and Meher (2019) mentioned that network of meteorological stations in the Himalayas is sparse with poor data quality containing huge missing values. According to the Ministry of Jal Shakti, prediction of such disasters is challenging primarily due to the lack of availability of sufficient data regarding rainfall intensity, location of a landslide, impounded volume and area, and physical conditions of lakes/water bodies. Therefore, glacial lakes and water bodies in the Himalayan region need to be closely monitored. Due to the continuous deposition of sediment in glacier lakes, temporal bathymetry data are required for close monitoring. To the author's knowledge, none of the studies related to GLOF were used remote sensing for glacier lake bathymetry estimation. However, remote sensing has been widely used in the estimation of submerged topographies of rivers. These techniques could be applied in the estimation of glacier lakes bathymetries.

3.2 Methods Used to Asses GLOFs

Most studies related to GLOFs assessment used remote sensing and GIS techniques. It includes glacial lake mapping using manual digitization, mapping based on spectral reflectance, land use cover, normalized difference vegetation index (NDVI), hillshade, slope, aspects, topographic position index, etc. Other studies used a modeling approach, hydro-meteorological time series data, and pixel-based time series data to assess GLOFs. Glacial lake bathymetries are generally measured from field surveys (as discussed above). Sometimes, it is not an easy task to do bathymetry surveys due to remote areas, logistic issues, and high cost. Alternatively, remote sensing can be used to estimate glacier lake bathymetries due to the frequent use of remote sensing in glacial lake studies. As the lake bathymetry is the most important input parameter in GLOFs modeling for flood risk and vulnerability, bathymetry needs to be estimated through remote sensing, especially in remote areas. Therefore, bathymetry estimation is important and relevant to GLOFs risk reduction. In this section, we will focus on the application of remote sensing in glacial lake bathymetry estimation.

3.2.1 Application of Remote Sensing in Bathymetry Estimation

Application of remote sensing in bathymetry estimation is generally based on the ratio of incident to reflected radiant flux which is called spectral reflectance. There are two general methods to drive bathymetry from satellite images or photographs (Hodúl et al. 2018; Cao et al. 2019).

3.2.2 General Approaches

A physical-based method relies on the radiative transfer models (RTM) where spectral reflectance is considered as a function of bottom reflectance, water depth, and water quality. Challenges in using the physical-based method are necessary atmospheric correction and accurate estimation of sea or river surface reflectance. Beer-Lambert law is used to measure water depth from brightness levels in imagery. Fir transparent medium of thickness x and a beam of light with incoming and outgoing intensities I_{in} and I_{out}, respectively can write as (Serway 1983):

$$I_{\text{out}} = I_{\text{in}} e^{-cx} \tag{1}$$

where c is the rate of absorption of the transparent medium. It will vary according to the turbidity of water and decrease exponentially as it passes through the medium. The terms I_{out} and I_{in} are final observed brightness level in imagery and initial brightness before the passage of light through the medium.

Empirical models rely on in situ bathymetry data where water depth drives from the relationship between in situ depth data and radiance of the image. In empirical models, water quality is assumed to be a constant.

3.2.3 Two-Media Photogrammetry Approach

Excluding these methods, some studies mentioned another approach based on the theory of two-media (earth's atmosphere and the water body of interest) photogrammetry based on Snell's law. The two-media photogrammetry includes four primary processes such as image extraction, noise and error removal, and refraction, and datum correction. In this approach, refracted rays in the water body were corrected by a refraction correction model to improve the accuracy of water depth estimation. Advantages of this approach over the previous two methods do not rely on accurate spectral information, not necessary to apply atmospheric corrections, suitable for remote areas and in situ sparse environments, and applicable in difficult atmospheric conditions.

There are three assumptions in the two-media photogrammetry method (Dietrich 2017; Hodúl et al. 2018; Cao et al. 2019). First is the planer water surface, uniform local water quality, and constant refractive index. The general methodology includes (1) calculation of the distance between the refraction points, (2) calculation of true water depth using observed water depth, and (3) calculation of the angle of incidence and refraction using observed water depth, the height of the flight, and horizontal distance between observed point and flight. Using these calculations, several errors are corrected. For instance, a horizontal error caused by approximate refraction correction model, a vertical error caused by average refraction angle, geometric model error, underwater mismatching, wave correction (either sea or river water), sea or river surface water elevation, and datum correction.

3.2.4 Passive Optical Remote Sensing Approach

Passive optical remote sensing approach for bathymetry estimation based on the spectral signature of a water column which is called spectral upwelling radiance. It depends on bottom albedo and water column optical properties. The optical properties of water vary with changes in suspended sediment and chlorophyll and organic matter content (Legleiter et al. 2004). Legleiter et al. (2004) mentioned a simple equation of spectral upwelling radiance as:

$$L_{\mathrm{T}} = L_{\mathrm{B}} + L_{\mathrm{C}} + L_{\mathrm{S}} + L_{\mathrm{P}} \tag{2}$$

where L_{B} is the portion of light that entered the water, L_{C} is the portion of the light that scattered into atmosphere, L_{S} is the portion of light that reflected from water surface without entering in it, and $\mathrm{L_P}$ is path radiance contributed by the atmosphere. Only L_{B} is related to water depth which is a function of both water depth and the light reflected from the streambed. Other studies including Legleiter et al. (2004) used complex and evaluated models for bathymetry depth estimation (Lyzenga 1978; Philpot 1989; Legleiter and Roberts 2005, 2009; Legleiter et al. 2018).

3.3 *Use of Photogrammetry in Bathymetry Estimation*

American Society for Photogrammetry and Remote Sensing (ASPRS), photogrammetry is the art, science, and technology of obtaining three-dimensional coordinates of physical objects and the environment through processes of recording, measuring, and interpreting photographic images and patterns of recorded radiant electromagnetic energy and other phenomena (Dewitt and Wolf 2000; McGlone 2004). A fundamental principle of photogrammetry is aerial triangulation which is a mathematical process used to establish an accurate relationship between the coordinate system of an image and the datum and projection (Schenk 1997). A centimeter-scale resolution can attain through color imagery which can measure the grain size of exposed areas (Carbonneau et al. 2004). A disadvantage of using color imagery of centimeter resolution is the requirement of thousands of images to cover a large area. Hence, variations in their spectral reflectance due to changes in camera exposure and aperture settings required high effort for their geometric and atmospheric corrections. However, for small areas, it would be useful.

Carbonneau et al. (2006) corrected variations in different color imageries using the calibration of color-depth relationships of feature-based image processing. They proposed a method capable of bathymetry depth estimation at 4 m^2 resolution with an error of ±15 cm. Fabris et al. (2010) generate DEM of an island's coastal areas with the help of in situ data and photogrammetric imageries with a vertical error of 1 m. Dietrich proposed an iterative approach for refraction correction of any camera combination to estimate bathymetry of an ideal condition with an average error of 0.02% of flight height. Shintani and Fonstad (2017) used orthophotographs (0.37 cm resolution) to estimate bathymetry with the help of a site-specific refractive index. The mean error of this study was 0.009 m with a standard deviation of 0.17 m. Kasvi et al. (2019) compared in situ, photogrammetry, and optical modeling results of shallow water bathymetry. They found that bathymetry estimation from photogrammetry is most affected by substrate variability and river bed material. Kaaminet al. (2020) used photogrammetric imageries for the identification of the boundary between land and sea with the help of simple image processing. Based on the literature, it is clear that photogrammetry is useful and high-resolution data to estimate shallow bathymetries in small areas.

Mostly glacier-related studies used photogrammetry in monitoring iceberg calving, change in glacier lake boundaries, and identification of glacier landforms (Šiljeg et al. 2015; Purdie et al. 2016; Kirchner et al. 2019; Fernández-Lozano and Andrés-Bercianos 2020). These studies used field survey instruments to measure glacier lake bathymetry.

3.4 *Use of Multispectral Imagery in Bathymetry Estimation*

Multispectral images capture the light of different wavelengths beyond the visible light range which can be combined to make a composite image. Composite images consist of several bands show different color patterns based on the spectral reflectance signature. The application of each band is different which is shown in Table 1.

Bathymetry retrieval from passive and multispectral imageries based on spectral reflectance is useful in homogenous water type and atmospheric conditions. For heterogeneous conditions, an independent source of information and algorithms are required along with spectral reflectance. Philpot (1989) mentioned a simple model for optically shallow water depending on wavelength in the homogenous condition

Table 1 Application of different bands in multispectral images (Weng 2010)

Bands	Wavelengths (μm)	Usage
Blue	0.45–0.52	For atmospheric and deep water imaging
Green	0.52–0.60	For imaging of vegetation and deep water structures
Red	0.63–0.69	For imaging of man-made objects, water up to 30 feet (9.1 m) deep, soil, and vegetation
Near infrared	0.76–0.90	For imaging of vegetation
Mid-infrared	1.55–1.75	For imaging vegetation and soil moisture content, geological features, and some forest fires
Thermal infrared	10.4–12.5	For imaging of geological structures, thermal differences in water currents, fires, and for night studies
Short infrared	2.08–2.35	Hydrothermal alteration minerals and mapping of alteration zones
Radar		For mapping terrain and for detecting various objects

is:

$$L_{\mathrm{d}} = L_{\mathrm{b}} e^{-gz} + L_{\mathrm{w}} \tag{3}$$

where L_{d} is radiance observed at the remote detector, g is an effective attenuation coefficient of the water, z is the depth of water column, L_{b} is a radiance term which is sensitive to bottom reflectance, and L_{w} is remotely observed radiance over optically deep water ($gz \rightarrow \infty$). Philpot (1989) also mentioned modifications of Eq. (3) for heterogeneous conditions which is used by recent studies (Lyzenga et al. 2006; Legleiter and Roberts 2009; Legleiter et al. 2018; Misra et al. 2018; Liu et al. 2020a, b). Band ratio analysis is also a widely used method for bathymetry retrieval. Legleiter et al. (2009) proposed that optimal band ratio analysis (OBRA) is a widely used model for bathymetry retrieval by a combination of band ratios. Niroumand-Jadidi et al. (2020) developed sample-specific multiple band ratio technique for satellite-derived bathymetry (SMART-SDB) which significantly improved the bathymetry retrieval relative to OBRA. It can be used for both inland and coastal waters.

3.5 *Use of Digital Elevation Model (DEM) Imagery in Bathymetry Estimation*

A digital elevation model is a representation of elevation values of a terrain. There are several types of DEMs with different resolutions. The publicly available advanced spaceborne thermal emission and reflection radiometer global digital elevation model (ASTER GDEM) is of 30 m resolution which is widely used to generate slope, contour, and drainage maps. It is generally used for watershed management studies. It can be used to estimate bathymetry for wide rivers (more than 30 m) and coastal areas. However, it has a vertical error which is reported in several studies (Pramanik et al. 2010; Gichamo et al. 2012).

The shuttle radar topography mission (SRTM) is of 90 m resolution also used widely for topography, slope, and elevation mapping. It is also used to estimate water depth in the river and coastal areas (Pramanik et al. 2010; Brêda et al. 2019).

Light detection and ranging (LIDAR) is another remote sensing technique to measure high-resolution (up to 1 m) elevation values of a terrain. Bathymetry LIDAR developed especially for bathymetry measurement mounted on a boat similar to sound navigation and ranging (SONAR). Topographic LIDAR measured elevation values up to water surface level. It cannot measure the submerged topography of a water body. Several studies combined LIDAR data with multispectral and photogrammetry imageries to improve bathymetry estimation.

Abdallah et al. (2012) compared ultraviolet and green LIDAR in bathymetry estimation and found 2.8 cm accuracy. Pan et al. (2015) tested topographic LIDAR in both fresh and turbid water. They found that green LIDAR performed well compared to others. Also, both water surface and turbidity influence the LIDAR performance.

Lague and Feldmann (2020) used green LIDAR to measure water depth which depends on water turbidity, bottom reflectance, and the sensor used. Tonina et al. (2020) used both bathymetry and topographic LIDAR to estimate the bathymetry of a gravel-bed mountain stream.

For glacier lakes, DEMs are generally used to map surrounding topographies, glacier lake water-level estimations, and glacier lake boundary monitoring.

3.6 Implication of Remote Sensing Application in Glacier Lake Bathymetry Estimation

Based on the literature, in the author's knowledge, none of the studies estimate bathymetry using remote sensing applications. Due to large lake areas, multispectral and DEMs would be useful to retrieve glacier lake bathymetry using a similar approach as in river and low-wave coastal areas. However, most of the remote sensing approach is for shallow water which reduces its ability in deep water lakes. With the help of low-density in situ data, a relationship between spectral reflectance or band ratio combination and in situ data could be developed in those areas where logistics are easily used. The developed relationship could be applied in glacier lakes of remote areas. Comparison of water turbidity and bottom reflectance of lakes and rivers would be an important factor affecting bathymetry estimation. Overall, a framework to execute the methodology of glacier lake bathymetry retrieval based on remote sensing applications in rivers and coastal areas needs to be developed. This framework would lie in between field surveys and remotely sensed data.

3.7 Economic implications on the Downstream Populations/sectors

Population living in the downstream area and depends on meltwater from glaciers and snow for drinking water, agriculture and irrigation, mining, and hydropower are affected during GLOFs. Motschmann et al. (2020) analyzed the effect of glacier lake due to climate change on human societies and found important negative but differentiated effects on natural and human systems. Shrestha et al. (2010) assess the downstream impact due to GLOFs in the Sun Koshi basin between Tibet (China) and Nepal. They found the nearly 900 households were affected which directly impacted a total population of 5800. The agricultural facilities and infrastructure of the area were vulnerable to GLOFs. The infrastructure includes roads, buildings, bridges, highways, dams, commination cables, water treatment plants, and powerhouses. The estimated total value of properties exposed to GLOF risk is US$159 million. Thompson et al. (2020) examine the institutional conditions under which Himalayan communities can create effective strategies to address GLOF risks.

4 Mitigation Measures

One of the most important risks in glacier lakes is the hydrostatic pressure of water on the moraine wall. The hydrostatic pressure can be reduced by lowering down the water level. There are four common methods for lowering down the water level such as the construction of an outlet control structure, pumping or siphoning out the water from the lake, boring a tunnel through the moraine barrier or under an ice dam, and controlled breaching (Bajracharya et al. 2007). Sharma et al. (2018) also mentioned these methods which were applied in the Nepalese Himalaya, for lowering down the water level up to 3 and 5 m. They mentioned that Siphons have a minimum risk of causing catastrophic failure and are suitable for installation in remote areas. Other mitigation methods are controlled deepening and widening of the outlet channel, construction of check dams in downstream areas to control outburst flow, and installation of an early warning system with a real-time satellite-based information system to minimize the GLOFs risk.

4.1 Case Study for Mitigation Measures

Nepal is one of the most vulnerable country in the world to the impact of climate change. Here, the average annual temperature is increasing with a rate of 0.04 °C and higher in the mountainous region. The temperature variation is increasing GLOF risk by contributing glacier contraction and lake expansion. As mentioned in Fig. 2 and according to the United Nation Development Program (UNDP), almost 24 active glacier lakes have been witnessed. This case study belongs to Solukhumbu district of Nepal that is also an active glacier lake. With collaboration of UNDP Nepal and Government of Nepal, a project was implemented to reduce to loss of life and property from GLOF in Solukhumbu districts and catastrophic downstream flooding. The main objectives of this project were to reduce GLOF risk and to reduce loss of life and property in some districts of the Terai area. Summary of achievable GLOF risk reduction is as follows:

- Successfully reduced water level by 3.4 m
- Sensors and early warning systems were installed in the high-risk zones
- Structures such as embankment were built to stabilize the course of the river
- Sediment trap measures were implemented
- High elevated tube wells and drainage system for flood-proofing were operationalized.
- Evacuation centers are developed
- Community-based early warning system was operationalized.
- Sediment monitoring protocols were developed
- A training manual was developed that trained several trainees
- Mock drill, street dramas, other awareness program were implemented.

Similar study was implemented in the Bhutan by UNDP with collaboration of the Royal Government of Bhutan. The focus of this study was on the most vulnerable Punakha-Wangdi and Chamkhar valleys to GLOFs. These valleys were included several basins and subbasins and lakes especially the Thorthormi lake. Thorthormi lake is considered one of the most dangerous glacial lakes of Bhutan that was outburst in the early 2010. The objective of this project was to reduce climate change-induced risks of GLOFs in the Punakha-Wangdi and Chamkhar valleys. Summary of the outcome of this project is as follows:

- Implementation of awareness programs in vulnerable communities to improve national, regional, and local capacities to reduce GLOFs risk
- Implementation of climate risk plans at district level
- GLOFs risk reduction through artificial lake-level management system
- Implementation of engineering and safety plans
- Proper water-level monitoring and maintenance
- Installation of early warning system to reduce loss of life and property
- Monitoring and maintenance of early warning systems
- Enhancement of learning, evaluation, and adaptive management.

5 Conclusion

The present review is based on the application of remote sensing in bathymetry estimation of the river and coastal areas and their scope in glacier lakes. In conclusion, the review reveals that most of flash floods in the Himalayas are due to extreme rainfall and climate change. Photogrammetry imageries are useful in small areas while multispectral imageries in large areas. Bottom albedo, optical properties of water, and water turbidity are important factors affecting spectral reflectance of multispectral imageries. Among DEMs, green LIDAR is effective in water depth retrieval. Due to the limited use of remote sensing applications in glacier lake bathymetry, a framework is required based on river and coastal water conditions. Furthermore, a relationship between bottom reflectance and water turbidity of the river and glacier lake water needs to be developed.

References

Abdallah H, Bailly JS, Baghdadi NN, Saint-Geours N, Fabre F (2012) Potential of space-borne LiDAR sensors for global bathymetry in coastal and inland waters. IEEE J Sel Top Appl Earth Observ Remote Sens 6(1):202–216

Allen SK, Rastner P, Arora M, Huggel C, Stoffel M (2016) Lake outburst and debris flow disaster at Kedarnath, June 2013: hydrometeorological triggering and topographic predisposition. Landslides 13(6):1479–1491

Bajracharya SR, Mool PK, Shrestha BR (2007) Impact of climate change on Himalayan glaciers and glacial lakes: case studies on GLOF and associated hazards in Nepal and Bhutan. International Centre for Integrated Mountain Development (ICIMOD)

Brêda JPLF, Paiva RCD, Bravo JM, Passaia OA, Moreira DM (2019) Assimilation of satellite altimetry data for effective river bathymetry. Water Resour Res 55(9):7441–7463

Cao B, Fang Y, Jiang Z, Gao L, Hu H (2019) Shallow water bathymetry from WorldView-2 stereo imagery using two-media photogrammetry. Eur J Remote Sens 52(1):506–521

Carbonneau PE, Lane SN, Bergeron NE (2004) Catchment-scale mapping of surface grain size in gravel-bed rivers using airborne digital imagery. Water Resour Res 40:7

Carbonneau PE, Lane SN, Bergeron N (2006) Feature based image processing methods applied to bathymetric measurements from airborne remote sensing in fluvial environments. Earth Surface Process Landforms: J Br Geomorphol Res Group 31(11):1413–1423

Das L, Meher JK (2019) Drivers of climate over the Western Himalayan region of India: a review. Earth-Sci Rev 198:102935

DeWitt BA, Wolf PR (2000) Elements of photogrammetry (with applications in GIS).

Dietrich JT (2017) Bathymetric structure-from-motion: extracting shallow stream bathymetry from multi-view stereo photogrammetry. Earth Surf Proc Land 42(2):355–364

Dubey S, Goyal MK (2020) Glacial lake outburst flood hazard, downstream impact, and risk over the Indian Himalayas. Water Resour Res 56(4): e2019WR026533

Fabris M, Baldi P, Anzidei M, Pesci A, Bortoluzzi G, Aliani S (2010) High resolution topographic model of Panarea Island by fusion of photogrammetric, lidar and bathymetric digital terrain models. Photogram Rec 25(132):382–401

Fernández-Lozano J, Andrés-Bercianos R (2020) On the origin of a remote mountainous natural reserve: insights from a topo-bathymetry reconstruction of the glacial lake of Truchillas (NW Spain). Quaternary International

Fushimi H, Ikegami K, Higuchi K, Shankar K (1985) Nepal case study: catastrophic floods. In: Young GJ (ed) Techniques for prediction of runoff from glacierized areas. Publication 149, International Association of Hydrological Sciences, Wallingford, pp 125–130

Galay V (1985) Glacier Lake Outburst flood (Jõkulhlaup) on the Bhote/Dudh Kosi-August 4

Gichamo TZ, Popescu I, Jonoski A, Solomatine D (2012) River cross-section extraction from the ASTER global DEM for flood modeling. Environ Model Softw 31:37–46

Hodúl M, Bird S, Knudby A, Chénier R (2018) Satellite derived photogrammetric bathymetry. ISPRS J Photogramm Remote Sens 142:268–277

Ives JD (1986) Glacial lake outburst floods and risk engineering in the Himalaya. Occasional Paper No. 5, International Center for Integrated Mountain Development, Kathmandu

Kaamin M, Fadzil MAFM, Razi MAM, Daud ME, Abdullah NH, Nor AHM, Ahmad NFA (2020) The shoreline bathymetry assessment using unmanned aerial vehicle (UAV) photogrammetry. J Phys: Conf Ser 1529(3):032109 (IOP Publishing)

Kasvi E, Salmela J, Lotsari E, Kumpula T, Lane SN (2019) Comparison of remote sensing based approaches for mapping bathymetry of shallow, clear water rivers. Geomorphology 333:180–197

Kanti T, Flemming G, Madhu J, Kuldeep J (2019) Extreme rainfall and vulnerability assessment: case study of Uttarakhand rivers. Nat Hazards 99(2):665–687

Kirchner N, Noormets R, Kuttenkeuler J, Erstorp ES, Holmlund ES, Rosqvist G, Holmlund P, Wennbom M, Karlin T (2019) High-resolution bathymetric mapping reveals subaqueous glacial landforms in the Arctic alpine lake Tarfala, Sweden. J Quater Sci 34(6):452–462

Kumar D, Singh AK, Singh DS (2020) Spatio-temporal fluctuations over Chorabari glacier, Garhwal Himalaya, India between 1976 and 2017. Quaternary International (in press)

Lague D, Feldmann B (2020) Topo-bathymetric airborne LiDAR for fluvial-geomorphology analysis. In: Developments in earth surface processes, vol 23. Elsevier, pp 25–54

Legleiter CJ, Roberts DA, Marcus WA, Fonstad MA (2004) Passive optical remote sensing of river channel morphology and in-stream habitat: physical basis and feasibility. Remote Sens Environ 93(4):493–510

Legleiter CJ, Roberts DA (2005) Effects of channel morphology and sensor spatial resolution on image-derived depth estimates. Remote Sens Environ 95(2):231–247

Legleiter CJ, Roberts DA (2009) A forward image model for passive optical remote sensing of river bathymetry. Remote Sens Environ 113(5):1025–1045

Legleiter CJ, Roberts DA, Lawrence RL (2009) Spectrally based remote sensing of river bathymetry. Earth Surf Proc Land 34(8):1039–1059

Legleiter CJ, Overstreet BT, Kinzel PJ (2018) Sampling strategies to improve passive optical remote sensing of river bathymetry. Remote Sens 10(6):935

Liu Y, Tang D, Deng R, Cao B, Chen Q, Zhang R, Qin Y, Zhang S (2020a) An adaptive blended algorithm approach for deriving bathymetry from multispectral imagery. IEEE J Sel Top Appl Earth Observ Remote Sens 14:801–817

Liu M, Chen N, Zhang Y, Deng M (2020b) Glacial lake inventory and lake outburst flood/debris flow hazard assessment after the Gorkha earthquake in the Bhote Koshi Basin. Water 12(2):464

Lyzenga DR (1978) Passive remote-sensing techniques for mapping water depth and bottom features. Appl Opt 17:379–383

Lyzenga DR, Malinas NP, Tanis FJ (2006) Multispectral bathymetry using a simple physically based algorithm. IEEE Trans Geosci Remote Sens 44(8):2251–2259

McGlone J (2004) Manual of photogrammetry, 5th edn. The American Society for Photogrammetry and Remote Sensing

Misra A, Vojinovic Z, Ramakrishnan B, Luijendijk A, Ranasinghe R (2018) Shallow water bathymetry mapping using Support Vector Machine (SVM) technique and multispectral imagery. Int J Remote Sens 39(13):4431–4450

Motschmann A, Huggel C, Carey M, Moulton H, Walker-Crawford N, Muñoz R (2020) Losses and damages connected to glacier retreat in the Cordillera Blanca, Peru. Clim Change 162(2):837–858

Muller F (1959) Eight months of glacier and soil research in the Everest region. Mountain World 1958/1959:191–208

Nandargi S, Dhar ON (2011) Extreme rainfall events over the Himalayas between 1871 and 2007. Hydrol Sci J 56(6):930–945

Niroumand-Jadidi M, Bovolo F, Bruzzone L (2020) SMART-SDB: sample-specific multiple band ratio technique for satellite-derived bathymetry. Remote Sens Environ 251:112091

Osti R, Egashira S (2009) Hydrodynamic characteristics of the Tam Pokhari Glacial Lake outburst flood in the Mt. Everest region, Nepal. Hydrol Proc Int J 23(20):2943–2955

Pan Z, Glennie C, Hartzell P, Fernandez-Diaz JC, Legleiter C, Overstreet B (2015) Performance assessment of high resolution airborne full waveform LiDAR for shallow river bathymetry. Remote Sensing 7(5):5133–5159

Philpot WD (1989) Bathymetric mapping with passive multispectral imagery. Appl Opt 28:1569–1578

Pramanik N, Panda RK, Sen D (2010) One dimensional hydrodynamic modeling of river flow using DEM extracted river cross-sections. Water Resour Manage 24(5):835–852

Purdie H, Bealing P, Tidey E, Gomez C, Harrison J (2016) Bathymetric evolution of Tasman Glacier terminal lake, New Zealand, as determined by remote surveying techniques. Glob Planet Change 147:1–11

Romshoo SA, Fayaz M, Meraj G, Bahuguna IM (2020) Satellite-observed glacier recession in the Kashmir Himalaya, India, from 1980 to 2018. Environ Monit Assess 192(9):1–17

Schenk T (1997) Towards automatic aerial triangulation. ISPRS J Photogramm Remote Sens 52(3):110–121

Serway RA (1983) Physics for scientists and engineers. CBS, pp 132–133

Šiljeg A, Lozić S, Šiljeg S (2015) A comparison of interpolation methods on the basis of data obtained from a bathymetric survey of Lake Vrana, Croatia. Hydrol Earth Syst Sci 19(8)

Sharma RK, Pradhan P, Sharma NP, Shrestha DG (2018) Remote sensing and in situ-based assessment of rapidly growing South Lhonak glacial lake in eastern Himalaya, India. Nat Hazards 93(1):393–409

Shintani C, Fonstad MA (2017) Comparing remote-sensing techniques collecting bathymetric data from a gravel-bed river. Int J Remote Sens 38(8–10):2883–2902

Shrestha AB, Eriksson M, Mool P, Ghimire P, Mishra B, Khanal NR (2010) Glacial lake outburst flood risk assessment of Sun Koshi basin, Nepal. Geom Nat Hazards Risk 1(2):157–169

Thompson I, Shrestha M, Chhetri N, Agusdinata DB (2020) An institutional analysis of glacial floods and disaster risk management in the Nepal Himalaya. Int J Disaster Risk Reduct 47:101567

Tonina D, McKean JA, Benjankar RM, Yager E, Carmichael RA, Chen Q, Leah CA, Kelsey G, Edmondson MR (2020) Evaluating the performance of topobathymetric LiDAR to support multi-dimensional flow modelling in a gravel-bed mountain stream. Earth Surf Proc Land 45(12):2850–2868

Veh G, Korup O, Roessner S, Walz A (2018) Detecting Himalayan glacial lake outburst floods from Landsat time series. Remote Sens Environ 207:84–97

Veh G, Korup O, von Specht S, Roessner S, Walz A (2019) Unchanged frequency of moraine-dammed glacial lake outburst floods in the Himalaya. Nat Clim Chang 9(5):379–383

Watanabe T, Rothacher D (1996) The 1994 Lugge Tsho glacial lake outburst flood, Bhutan Himalaya. Mountain Res Dev 16(1):77–81

Weng Q (2010) Remote sensing and GIS integration: theories, methods, and applications. McGraw-Hill, New York

Impact of Floods on the Green Energy Sector in Himalayas—A Case Study of Gagas Watershed, Uttarakhand, India

Sapna Bisht, Smita Chaudhry, Subrat Sharma, and Surajit Dutta

Abstract There is a striking absence of knowledge about the characterization and vulnerability of energy resources such as small-scale hydropower in the face of environmental changes. There is a need to understand water resource sustainability, extreme events, land use changes, etc. as well as their social contestations in the watersheds. Based on these parameters, this study was carried out in the Gagas watershed (Western Himalaya). The sustainability of watermills as small-scale hydropower units was analysed on the basis of field data and remote sensing and geographic information system data. Around 12% of watermills were observed to be functional and 88% were non-functional. In the land-use/land-cover change analysis of 38 years, forest had increased (7.71%) and sown area had decreased (0.50%) in the study area. The primary reasons for non-functionality were observed to be reduced profitability and increased maintenance costs, majorly on the components. The Principal component Analysis (PCA) determined supportive villagers (41%) and lack of managerial sustenance (17%) as major functionality factors for small-scale hydropower units. The frequent occurrence of extreme events such as floods, and the fact that 31% of the total area is within very-high to high flood susceptibility zone, affects the expenditures on maintenance. This study can be useful for policy development, technological up-gradation and sustainable development and in the better planning of renewable resources.

S. Bisht (✉)
International Crops Research Institute for the Semi-Arid Tropics (ICRISAT), Patancheru, Hyderabad, Telangana 502324, India

S. Chaudhry
Institute of Environmental Studies, Kurukshetra University, Kurukshetra, Haryana 136119, India

S. Sharma
Ladakh Regional Centre, G.B. Pant 'National Institute of Himalayan Environment' (NIHE), Wildlife Warden Office Building, Near Council Secretariat, Leh, Ladakh(UT) 194101, India

S. Dutta
Department of Geography, Dr. Shyama Prasad Mukherjee University, Ranchi, Jharkhand 834008, India

S. Kanga et al. (eds.), *Disaster Management in the Complex Himalayan Terrains*, Geography of the Physical Environment,
https://doi.org/10.1007/978-3-030-89308-8_8

Keywords The himalayan mountains · Rural energy · Small scale hydropower · Green energy

1 Introduction

In a developing country, such as India with its increasing energy demand, with increased consumption and increasing gaseous emissions, it has become imperative to develop and promote alternative and sustainable energy sources (Bisht and Sharma 2018). The country has varying altitudinal ranges that are transversed by numerous streams, canals, and rivulets. The Indian Himalayan range (IHR), in particular, has great potential for harvesting hydropower for generating mechanical and electrical power owing to being a home to traditional technologies developed in local environments. One among these technologies is the traditional watermills (*gharats*) used for Mechanical power generation, which has served throughout the history of settlements in the mountains. In the mountainous regions, rural people have historically met their energy needs by using biomass, human labor, imported kerosene, and traditional water powered vertical axis mills. Owing to the fact that the availability of different types of energy resources plays a very important role in determining energy security or energy scarcity, and the choice may lead to various stages of environmental degradation.

In the Himalayan region, studies have revealed that nearly 500,000 watermills, spread in the entire region, can produce as much as 2500 MW of power (assuming each generates 5 kW) along with a cash generation of Rs. 1200 million per hour (Prasad and Prasad 2007; Paul and Khan 2010); giving a new direction to social wellbeing of the society in terms of giving employment to millions of people and as an in-expensive and established technology (Dotti and Misra 1987; Sengupta 1987; Saini and Kumar 2006). But the characterization and vulnerability of such small-scale hydropower generations in the face of environmental changes, viz., water sustainability, and extreme events coupled with human activities such as increased water consumption and land use changes in their watersheds are questionable. In India, small-scale hydropower research has concentrated more on policy, financial and technical conditions, and contexts. There is a striking absence of what these small-scale units were actually doing in their social contestation and acceptance. In this study, it underlines the importance of studying such potential green-energy in their local context. Therefore, the objective of this study was to analyse small scale hydropower, i.e., watermills in the context of hydro-ecological, economic, and social sustainability.

1.1 Impact of Floods in the Himalayas, Specifically Sudden Storm Events, on Different Sectors

Extreme weather events like increased precipitation, cloudbursts, flashfloods, and avalanches -events that occur with little warning in communities in the mountainous Hindu Kush Himalayan (HKH) region are particularly challenged by this, threatening lives, livelihoods, and state, and national economies. Floods accounted for 35% of natural disasters during 1975–2005 in South Asia (Shrestha and Takara 2008). Vulnerable groups such as the poor, women, children, the elderly, and people with disabilities are often the hardest hit. The combination of unstable atmospheric, cryospheric and geologic processes, as well as highly favorable steep topography, creates ideal conditions for such hazards resulting in fatalities, displacement of people, and damage to millions of dollars of property annually in the HKH (Meraj et al. 2015; Dimri et al. 2016; Meraj et al. 2018a, b; Kumar et al. 2018a, b; Aryal et al. 2020). Although the Himalaya is highly vulnerable to all types of water-induced hazards such as flash floods, soil erosion, landslides, glacial-lake outburst, and debris flow in particular, in monsoon period and drought in non-monsoon period (as drying up of natural water springs and streams) (Meraj et al. 2016; Rather et al. 2017). However, in recent years, flash floods and river line floods caused by torrential rains or localised convective events have been observed (Altaf et al. 2013; Rawat et al. 2012; Arora et al. 2016). It has been known as one of the extensive ongoing environmental issues that have wide impacts on agriculture, ecology, hydrology, and in socioeconomic contexts (Nandargi and Dhar 2011; Dewan 2015; Bhatt et al. 2017). Relating to the intricate linkages between ecology, economics, and vulnerability in the Himalayan ecosystem, Maikhuri et al. (2017) and Bell et al. (2021) have shown that the Himalayan floods have tremendously impacted the following:

1. Livelihoods such as • Pilgrimage tourism • Agricultural crops and fruit trees • Livestock • Forest/alpine meadows • Climate sensitive natural resources-based livelihoods.
2. Health-related impacts such as • Increase in the number of patients (psychological trauma) • Depression, hypertension, and other diseases.
3. Education-related consequences • Weakening Education systems (less enrollment and dropout) • Increased inaccessibility to primary/secondary schools.
4. Infrastructural-related impacts • Roads, motor and bridle bridges, schools, and health centers destroyed/damaged. • Damage to houses, schools, buildings • Cultivated fields destroyed.
5. Economic activities • power generation disrupted due to damage to the hydropower stations.

1.2 *Impact of Floods on the Green Energy Sector – Small Hydropower Projects*

Flash floods and cloudbursts cause massive damage in downstream regions where the floods destroy structures, cause siltation in rivers, downstream reservoirs, dams, and lakes, and on several occasions, caused fatalities (Prasad and Roy 2005; Mazari and Sah 2005). As stated earlier, in the fragile mountain areas like the Himalaya, such calamities, in their majority, are associated with different forms of landslides, land-slips, and soil erosion. As the frequency of natural disasters in the region increases, so do the destructions and casualities. Studies have implicated the reason as the constructions of hydropower projects on larger scales (Sharma et al. 2007). The establishment of hydropower projects and the occurrence of flash floods follow a circular path while one contributes to the other. It is pertinent to notice here that while a dry season reduces production, a wet season might not increase the electricity production (Alley et al. 2014). In the small run-of-river hydropower plants with no impoundment, the mountainous streams with frequent flash floods carrying high sediment loads causes high damage to the turbines and leads to a drop in hydraulic efficiency (Weerakoon and Rathnayake 2007; Singh et al. 2007).

Sustainability assessment studies have shown that hydropower projects in the capacity range of 1–5 MW are the most sustainable (Kumar and Katoch 2015). However, over the years, increasing uncertainties about river flows due to climate variability have raised a major concern about future projections of hydropower production, infrastructure design, flood control, and their associated risk assessment and planning (Ahlers et al. 2014; Schwanghart et al. 2016; Agarwal and Kansal 2020; Talchabhadel et al. 2021). In the HKH region, without adequate planning, the majority of hydropower plants are built in locations that are extremely vulnerable to hazards such as landslides, earthquakes, heavy siltation, and flash floods (Sahu et al. 2020).

2 Case Study—Gagas Watershed Hydropower Vulnerability to Natural Hazards

The study area, Gagas watershed, is situated in Almora District, Uttarakhand, between 29°51′55″ N and 29°35′49″ N latitude and between 79°20′36″ and 79°33′15″ E longitude in the Western Himalaya. It is a part of a larger Ramganga basin (Fig. 1).

The geographical area of the watershed is approximately 511.19 km^2. Although in the mountains, the distribution of rainfall is affected by the topography, i.e., altitude variations, slope, aspect, and trends of mountain ranges, for a general overview, the analysis of climatic data for 50 years (years 1950–2000) showed that the maximum average rainfall occurred in July and the minimum in November. The monthly variations of the climatic data are shown with the help of hyetographs and temperature isotherms (Fig. 2).

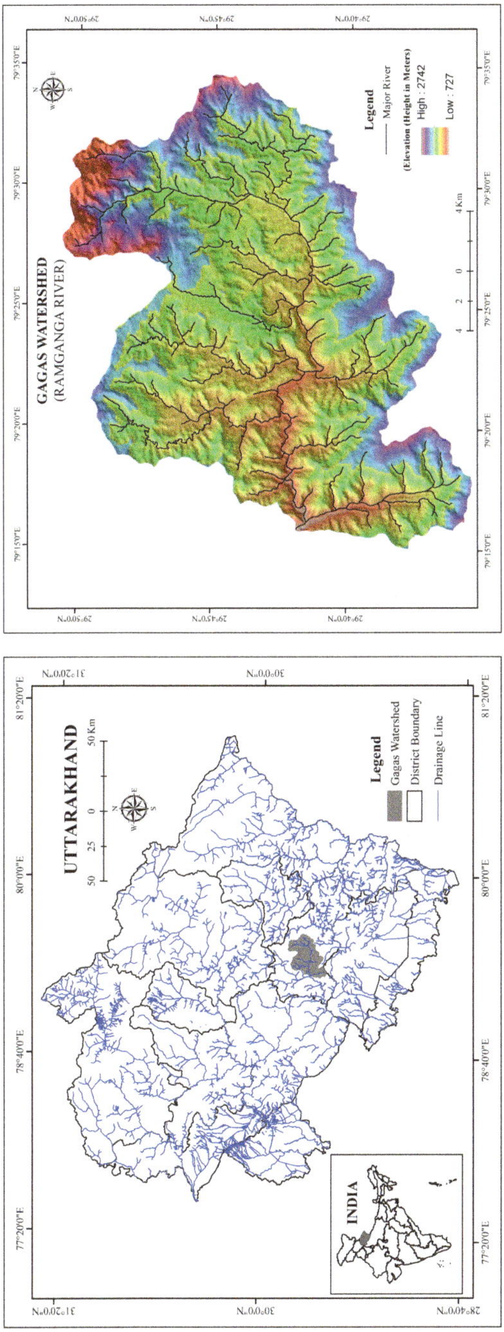

Fig. 1 Study area of Gagas watershed in Almora district of Uttarakhand

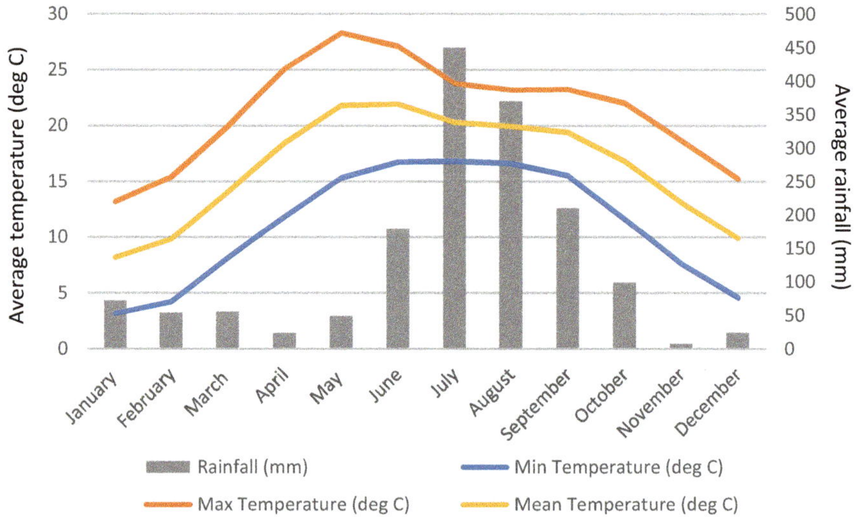

Fig. 2 Monthly climatic details of the study area (years 1950–2000). *Source* Bioclim methodology

For studying hydro-ecological sustainability of the watermills, the spatial distribution using land use/land cover change analysis and the hydro-meteorological conditions were analysed in the study area. For land use/land cover change (LULC), the main data sources were toposheets from Survey of India (SOI) toposheets (1:50,000 scale), the Landsat 8 operational land imager and thermal infrared sensor (OLI/TIR), and Landsat multispectral scanner system (MSS) satellite imageries which were collected from the USGS Website. The ground verification was carried out by GPS Garmin 72. Land use/land cover maps were prepared in Arc GIS 9.3 and ERDAS 9.1 software (Table 1).

Spatial and temporal LULC analysis of the study area was carried out using two satellite images, namely Landsat MSS and Landsat 8 OLI/TIR ETM+. The

Table 1 Data and specifications of SOI toposheets and satellite imageries

Study area	Satellite sensor/toposheet number	Path/row	Date of acquisition	Spatial resolution (m)
Gagas (Uttarakhand)	Landsat 8 Landsat MSS	Path 145 and Row 39	21.02.2015 17.02.1977	30 79
	530/5, 530/6, 530/9, 530/10 Scale: 1:50,000	–	1958	
	ASTER GDEM2	Path 145 and Row 39		30

imageries were geo-referenced using toposheets and the nearest neighborhood algorithm method. The satellite images were radiometrically corrected for referencing, and the study area was subset. After preparation of the initial map, field surveys were carried out with the help of pre-map and hand-held GPS. On the basis of classified polygons, the unmatched classes were marked on the map and cross checked.

Based on the field survey, eight major LULC classes, namely current fallow land, urban settlements, forest, barren/wasteland, scrub/degraded forest, river channel, sown areas, and village settlements were considered for LULC classification and change detection in the study area (Nathawat et al. 2010). Forest land has been considered as an area with not less than 0.5 hectares of trees and canopy cover of trees more than 10% (forest resources assessment 2015). Sown area was considered as an area under the current crop, i.e., wheat. The land use/land cover classification was done with the help of on-screen visual interpretation with the help of interpretation keys. After preparation of the initial map, a field survey was carried out with the help of pre-land use/land cover map and GPS. The selection of ground truth 110 points in Gagas were collected using a stratified random sampling method.

For the analysis of extreme rainfall data, the daily rainfall data from APHRODITE was analysed using ArcGIS 9.3 software. These daily rainfall data was analysed by calculating the total number of wet days and the total number of heavy rainy days for each year. In terms of daily rainfall analysis, greater and equal to 10 mm of rainfall was taken as heavy rainfall (Balling et al. 2016).

For studying the economic and social sustainability of watermills, a total of 77 watermills were studied in different sub-watersheds. Various possible reasons for success and failure and the degree of impact from each of them were quantified using a questionnaire method. "Factor analysis" was used to reduce the major factors using the SPSS tool. The respondents rated each variable on a scale of 1 (very low)—5 (very high). Figure 3 shows a broad framework of the methodology adopted.

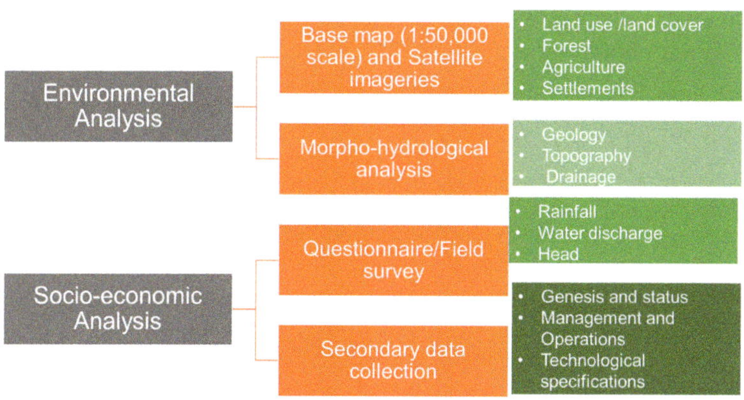

Fig. 3 Comprehensive methodology for analysing sustainability factors

3 Results and Discussion

In the Gagas watershed, environmental degradation has led to a significant decline in the quality of life. The entire watershed had observed a tremendous change over the last 50 years. The analysis of sustainability parameters was carried out and is discussed as below.

3.1 Hydro-Ecological Sustainability

In the watershed, the most effective precipitation events were largely concentrated in the monsoon months, as shown in Fig. 4. In the monsoon season, the accelerated overland flows caused devastating floods and brought along dissolved and suspended materials and bed load (Joy et al. 2019). Annual precipitation ranged from 318 to 988 mm, with average rainfall ranging from 30.5 to 85.4 mm, minimum number of rainy days were 31 days, and the maximum number were 75 days in different sub watersheds. The watershed experienced occasional flash floods (Kumar and Kumar 2008).

Evidence has emerged that climate change has made the variability more intense, with an increased frequency of extreme events such as droughts, floods, and hurricanes (Gujree et al. 2017). In the analysis of the number of days of heavy rainfall in the Gagas river watershed, it showed that there was an increase during 1980s to 2007 as compared to the duration of the 1950s to 1970s. It was mainly because of the changes in the climate pattern. Days of heavy rainfall were observed to be at their maximum in the year 1991 in the Gagas river watershed (Fig. 5).

In the analysis of the number of rainy days in the Gagas river watershed, the value of R2 represented a linear relationship in rainfall. The rate of change of slope was observed to be negative, representing that the rate of change of rainy days had decreased in the duration from the 1950s to 2007 (Fig. 5).

LULC was analysed to study the linkages and feedbacks regarding climate change, other social, economic and environmental aspects, and their impacts (Fig. 6).

The LULC maps for the duration from 1977 to 2015 showed considerable change during the 38 year duration. In the watershed, the forest cover had increased by 2.8%, majorly due to afforestation programmes by the government as well as the NGOs working in the area. Urban settlements, sown areas, and rural settlements, all increased by 0.2%, 0.5%, and 0.1%, respectively. The sown area represented the agricultural land for rabi crops which included wheat, barley etc., and were the major portion of grains brought to the watermills for milling purposes.

The transition matrix from the duration of year 1977 to year 2015 was analysed to study the land use conversions (Table 2) and provided essential data on the nature and spatial distribution of land use changes (Kanga et al. 2017; Kumar et al. 2018a, b). During the study period, forest cover had increased and gained area from other land use categories. The forest covered 199.8 km^2 in the year 1977 and 214.4 km^2

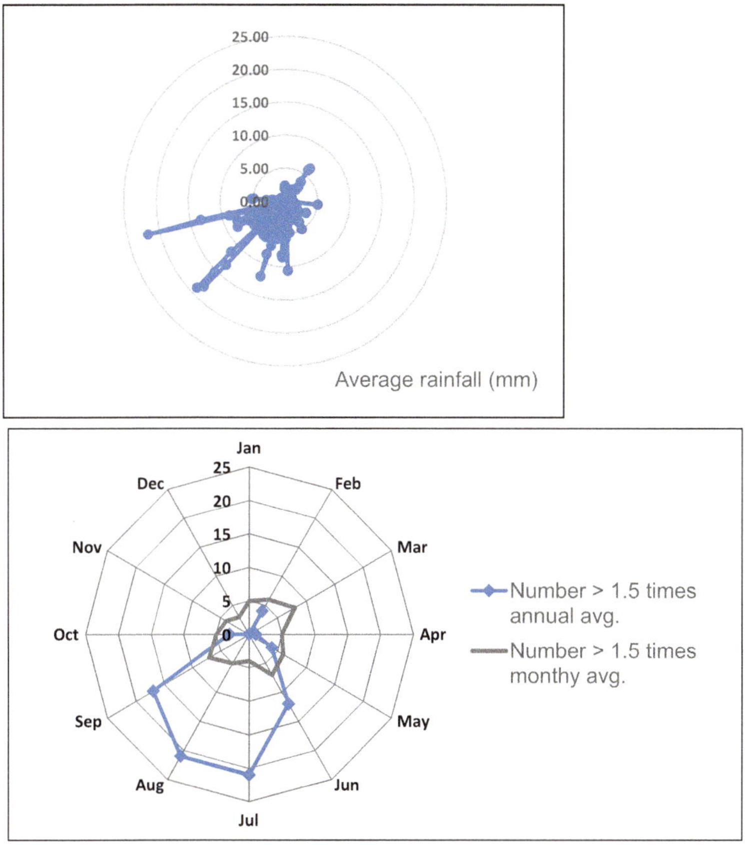

Fig. 4 The hydrometeorological situation in the Gagas watershed is as follows: **a** average precipitation (mm) from 8 meteorological stations over a 3 year period: 2009–2011; **b** The number of precipitations on each day of the year that exceeds 1.5 times the annual and monthly averages for the period from 2009 to 2011

in the year 2015. The increase in forest area from 1977 to 2015 was 35.8 km^2 from barren/wasteland, 27.7 km^2 from the current fallow. The area under the sown areas had also increased from 14 km^2 in the year 1977 to 16.4 km^2 in the year 2015. The major increase was from current fallow, i.e., 5.6 km^2. The urban area also increased during the years 1977 to 2015, with a 0.2% increase. In the study area, there was a decrease in current fallow, i.e., from 133.6 to 129.8 km^2. The major portion of it was diverted to forest and barren/waste land.

The land use/land cover study signifies the linkages in the upstream–downstream physical environment. The linkages are complex due to the extreme altitudinal range associated with the young and fragile geology, seasonal and spatial variation in rainfall, and the diversity of anthropogenic processes. Land use changes have the potential to impact the hydrological regime and influence the downstream water

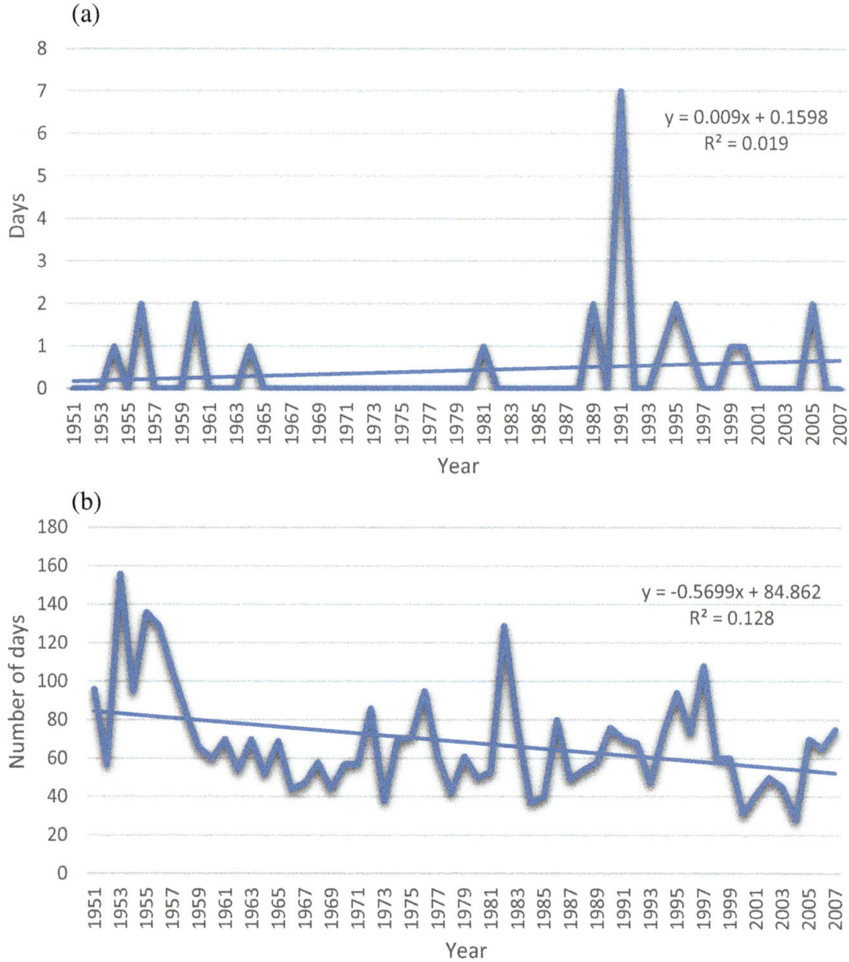

Fig. 5 **a** Heavy rainfall; **b** Number of rainy days in the Gagas river watershed

availability (Nepal et al. 2014). Geo-environmentally, based on the rate of runoff, sediment load delivery, and denudation during the rainy season, barren lands (i.e., riverbeds, barren/waste land, scrub/degraded etc.), as well as agricultural lands, have extreme vulnerability to flood hazard, whereas, dense forest areas (i.e., oak, pine and mixed forests) have low vulnerability (Rawat et al. 2012).

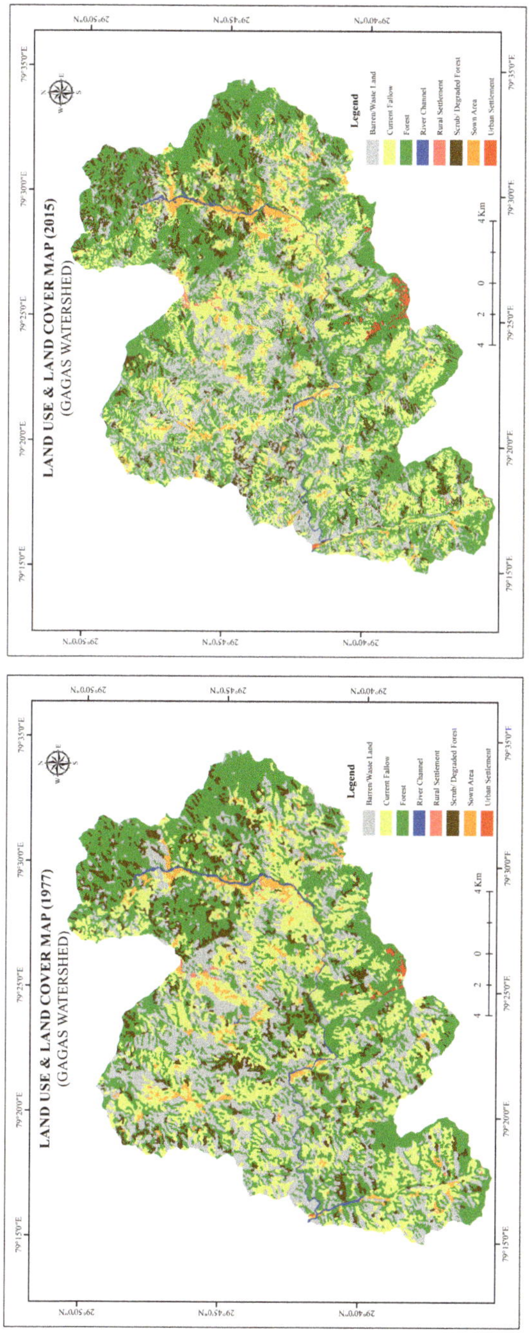

Fig. 6 Land use/land cover maps of the year 1977 and year 2015

Table 2 Land use/land cover change matrix between the years 1977 and 2015 (km^2)

1977	2015								
	Current fallow	Urban settlement	Forest	Barren/waste land	Scrub/degraded	River channel	Sown area	Rural settlement	Total
Current fallow	**75.8**	0.5	27.7	19.4	3.9	0.2	5.6	0.5	133.6
Urban settlement	0.0	**0.8**	0.2	0.0	0.0	0.0	0.0	0.0	1.0
Forest	26.2	0.7	**132.9**	24.9	12.3	0.2	2.3	0.2	199.8
Barren/waste land	17.5	0.0	35.8	**54.7**	9.4	0.2	1.5	0.1	119.3
Scrub/degraded	4.7	0.0	15.5	10.0	**7.6**	0.0	0.5	0.0	38.3
River channel	0.6	0.0	0.5	0.6	0.1	**1.8**	0.3	0.0	3.9
Sown area	4.6	0.1	1.6	0.9	0.4	0.2	**6.1**	0.2	14.0
Rural settlement	0.2	0.0	0.1	0.0	0.0	0.0	0.1	**0.8**	1.2
Total	129.8	2.0	214.4	110.5	33.7	2.7	16.4	1.8	**511.2**

3.2 Techno-Environmental Sustainability

As the most determining factors in the functionality of these watermills, their stream discharges and the available heads in the streams/'gadheras' were observed and analyzed. The head range of studied watermills ($n = 77$) ranged from 2.5–3.6 m and water discharge ranged from 0 to 0.95 cumec. The graphs showed that these two parameters were not the only determining factors for the functioning of watermill units in the studied area. Both these factors were found to be lower in functional watermills as compared to nonfunctional watermills (Fig. 7).

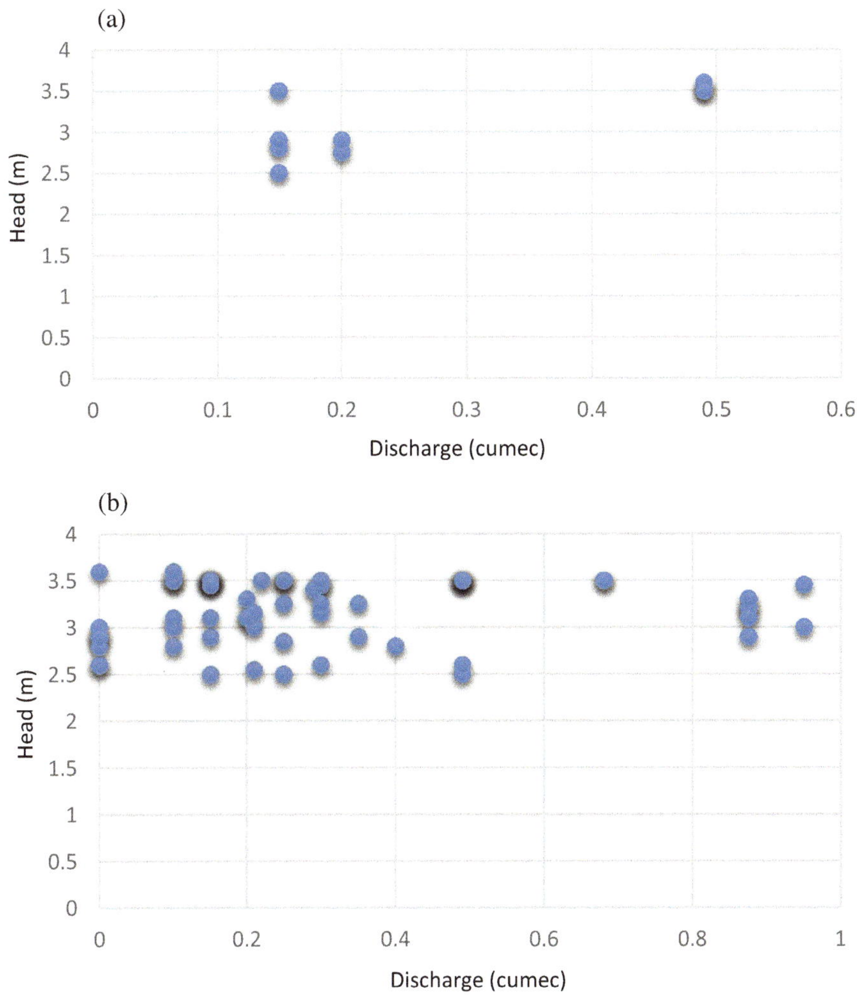

Fig. 7 Graphs showing the head and discharge of watermills. **a** Functional ($n = 9$) and **b** non-functional ($n = 68$) watermills

3.3 Economic Sustainability

Considering the fact that in the study area, the owners of watermills largely depended on local resources for making various components of watermills, viz.,

1. The milling stones used for grinding grains were brought from selective locations in the region. Earlier, a pair of these used to be brought by 4–8 people due to lack of road networks and/or commutation facilities. The cost of these during earlier times was observed to be Rs. 262 ± 237, which has now increased to Rs. 4,385 ± 2,694. The age of the stones generally depended upon the width of the stones (6–12 inches) and how frequently they were used. The age was observed to be around years 7.28 ± 6.18 with an average loss of 1 inches/year. The masons who extracted/prepared these were observed to be less available now.
2. Pipes/chutes used to convey water from a required height on to the runner to rotate it and were made up of wood, majorly *Pine spp.* from local forests. The costs of preparing a chute were around Rs. 577 ± 451 and it lasts to an age of years 7.78 ± 6.48.
3. The runners were made from forest wood and cost around Rs. 325 ± 41.8, and their shelf life limit were about years 10.1 ± 10.2. The age mostly depended on which wood these were prepared from and how they were maintained, especially in dry seasons. To last a long time, some wood requires constant submersion in water, such as *Bombax ceiba.* The most sought-after tree/log wood for long-term use was *Pine spp.* wood with gum/resin. As the gum protected the wood from being rotten by water.
4. The hoppers were prepared only once in their life time, generally by using bamboo or *Pine tree spp.* or by using tin. It cost about Rs. 50.
5. There were several other parts of the traditional watermills, such as bearings (pivots), which were earlier brought from the Pindar valley and have now been replaced by iron. Other smaller parts included wooden blades, base plates, wooden bushes, lifting and lowering systems, etc. These were prepared using wood from the forest itself. Different types of forest tree species were utilised for making such components such as *Pinus roxburghii*, *Bombax ceiba*, *Myrica esculenta*, *Sapium insigne*, *Toona serrata*, *Rhododendron arboreum*, *Quercus leucotrichophora*, and bamboo based on their availability in the local environment (Table 3).

As a major portion of the grains brought for milling cost benefit, analysis of the wheat (*Triticum aestivum*) was carried out. It was observed that, since the earlier times, the rate of milling was not changed, i.e., 0.79 ± 0.27 per 10 kg or Rs. 1.42 ± 1.01 per kg of wheat milled. Finger millet (*Eleusine coracana*) was the second most common grain brought for milling, followed by maize (*Zea mays*), barley (*Hordeum vulgare*), black gram (*Vigna mungo)*, sorghum (*Sorghum bicolor*), barnyard millet (*Echinochloa frumentacea*), rice (*Oryza sativa*), and spices such as turmeric (*Curcuma longa*). In terms of benefit, it was observed that the mill

Table 3 Specifications of the components of watermills and their sources

Components name	No. of respondents (*n*)	Sources	Cost (Rs.)	Age of function (yrs.)
Milling stone	40	Selected locations in the region	Present: 4385 ± 2694 Earlier: 262 ± 237 Reused: 1267 ± 503	7.28 ± 6.18
Pipe/chute	38	Forest	577 ± 451	7.78 ± 6.48
Runner	28	Forest	325 ± 41.8	10.1 ± 10.2
Hopper	12	Market, forest	50	life time
Pivot	5	Riverine/market		
Wooden blades	10	Forest	367 ± 153	1.7 ± 1.7
Base plate, wooden bush, lifting and lowering system	5	Forest		

owners earned more profit earlier. The reason behind this was that more people were used to visiting the watermills for milling wheat grains, and now villagers bought them from the market in the form of wheat flour. Subsequently, the earnings now have decreased from 182 ± 177 kg/month to 81 ± 78 kg/month. The expenditures were majorly observed to be on maintenance of components, i.e., Rs. 5704 ± 3340 (per app. 7 years). There were other recurring expenditures on the constructions of dams (temporary), i.e., Rs. 1940 ± 669, and canals, i.e., Rs. 250 ± 132. Some of the owners renovated their units' sheds and dams permanently and thus the increased costs. Overall, the benefits from the units had decreased over the years and expenditure had increased (Table 4).

Seasonality of watermills affected the financial output for instance, among the functional watermills, 78% were only seasonally functional and were thus closed in summers due to lack of water in streams and only 22% were in working condition in all seasons. The other factors also determined functionality as well as the benefits of watermills, such as the maximum number of functional watermills, which had minimum distances from nearest villages, road heads, the owner's home, and markets from villages compared to nonfunctional watermills.

3.4 Social Sustainability

In the studied watermill units, out of the total watermills, 12% were in working conditions and 88% of the watermills were defunct. Generally, human behaviour and psychological factors act as barriers to rural social change in the study area.

Table 4 Cost benefits analysis of watermills

	Rate of milling (Benefit)		Earnings (Kg/m)		Expenditure on components (Rs. per app. 7 years)	Recurring expenditure (Rs./yr)		Other expenses (one time (Rs.))	
						Dam (stone)	Canal	Cemented shed	Cemented dam
	kg/10 kg	Rs./kg	Earlier	Present					
Average	0.79	1.42	182	81	5,704	1940	250	32,500	22,100
SD (±)	0.27	1.01	177	78	3,340	669	132	24,749	18,258

Table 5 Major factors responsible for watermills failures (using PCA analysis)

Factors	Variables	% Responsible for failure
Managerial sustenance	Water usage for different purposes	16.71
	Noninterest	
	Miscellaneous	
Migration issues	Unskilled Operator	12.69
	Non profitable	
Financial issues	Financial dispute among partners	10.45
	Misuse by villagers	
	Lack of funds for maintenance and repair	
Non efficiency	Options of alternate energy sources	8.77
	Time consuming (negative relation)	
	Low agriculture production	
Natural calamities	Natural calamities (negative relation)	8.42
Technical issues	Technical difficulties	6.79

In the study area, it was observed that an important feature of most of the working watermills was that these were individually owned rather than community owned. The decreased chances of disputes and the ability to better manage them may be the observed reasons. A principal component analysis (PCA) was performed on the correlation matrix among the different parameters followed by Varimax rotation with Kaiser normalization, with the same being used to examine the association between them. The factor loadings were classified as "strong," "moderate," and "weak" corresponding to absolute loading values of >0.75, 0.75–0.50, and 0.50–0.30, respectively (Liu et al. 2003).

For determining factors for the failure, the parameters were loaded for six components from the PCA of the dataset. The table shown below gives the detailed composition of each factor extracted with the percentage of variance explained by it (Table 5). The first factor explains the maximum variation of 16.7% of the total. This was represented by positive strong loading for "water usage for different purposes" and positive moderate loading for "noninterest" and positive strong loading for "miscellaneous" that included subfactors such as old age, river changing directions, drought, death, or noninterest, and these variables were considered as the factors describing managerial sustenance. Factor two accounted for 12.7% of variation, with highly positively loadings for “unskilled operator” and “nonprofitable.” The third factor explained 10.45% of variation and showed positive moderate loading for “financial dispute among partners” and “misuse by villagers” and “lack of funds for maintenance and repair.” The fourth component accounted for 8.77% of the variation and was represented by moderate loading for “options of alternate energy sources,” “low agriculture production,” and negative moderate loading for “time consumption.” The fifth component explained 8.4% of variation and showed negative loading for “natural calamities.” The sixth factor explained 6.8% of variation and showed strong positive loading for “technical difficulties.”

The analysis of the factors contributing to success of watermills in the Gagas watershed led to the extraction of five principal components. The first factor explained the maximum variation of 40.97% of total and was represented by positive moderate loading for “additional income generation” and strong loading for “cooperative villagers” and “supportive co-partners.” These variables can be considered as the factors describing supportive villagers. Factor two accounted for 16.51% of the variation and showed moderate positive loading for “additional income generation” and “financial stability” and negative strong loading for “experienced operator” and “institutional support.” The third factor accounted for 14.74% of the loading and showed strong positive loading for “efficient management,” “proper seasonal maintenance” and moderate negative loading for “institutional support.” The fourth component explained 11.32% of the variance and was represented by moderate loading for “old custom,” strong loading for “availability of perennial source” and strong negative loading for “livelihood source.” The fifth component explained 9.06% of the variance and represented strong positive loading for “availability of equipment.” The main factors responsible for the functionality of watermills can be categorized as: supportive villagers, financial stability, effective management, traditional beliefs, and availability of equipment (Table 6).

3.5 *Flood Vulnerability*

For the assessment of flood risk vulnerability in the watershed, relevant parameters such as stream properties, basin characteristics, and maximum stream discharge were analysed using RS/GIS in the published study by Bisht et al. 2016. Ranking according to compound value represents the flash flood susceptibility for different sub-watersheds. After analysis of compound values, it was observed that an area of 137 km^2 which is 31% of the total area, was of very high priority and very highly susceptible to flash flood occurrences, an area of 246 km^2 which is 56% of the total area, was of moderate priority for flash flood susceptibility.

Table 6 Factors responsible for functionality (using PCA analysis)

Factors	Variables	% Responsible for success
Supportive villagers	Additional income generation	40.97
	Cooperative villagers	
	Supportive co-partners	
Financial stability	Additional income generation	16.51
	Experienced operator (negative relation)	
	Financial stability	
	Institutional support (negative relation)	
Effective management	Efficient management	14.74
	Proper seasonal maintenance	
	Institutional support (negative relation)	
Traditional beliefs	Livelihood source (negative relation)	11.32
	Old custom	
	Availability of perennial source	
Equipment available	Availability of equipment	9.06

4 Conclusion

In mountainous societies, as the most vulnerable groups in the wake of climatic change, the study of sustainable energy resources may be considered as important and crucial. To avoid depletion of natural resources and maintain an ecological balance, all the elements of sustainability, i.e., environment, society, and economy have been analyzed. Climatologically, in the study area, the frequency of heavy rainfall has increased in recent years, whereas the number of rainy days has decreased during the period, owing to increased vulnerability to such extreme events and a dearth of water availability in other seasons. The changes in the local resources were analysed using the LULC change matrix for a period of 38 years. The analysis showed an increase in sown agricultural land and forests in the last four decades.

Socioeconomically, the study of functionality factors was analysed using PCA. The result showed that the major success factors were active villagers and financial stability, while major failure factors were financial and management sustenance issues and disinterested villagers. High dependability on natural resources was observed in the watershed. Moreover, in the cost-benefit analysis, the expenditures on maintenance were observed to be higher as compared to profit. An area of 137 km^2, which is 31% of the total area, was in very high priority and very highly susceptible to flash flood occurrences. In the entire watershed, more than 85% of the total area was found to be highly vulnerable to flash floods. Changing climatic conditions such as extreme weather conditions, floods, and low seasonal availability of water has increased the maintenance costs of these hydro units.

However, these small scale hydropower units help in CO_2 mitigation by restricting the use of diesel in diesel mills and saving electricity consumption in electricity mills. There are various shortcomings and factors that need to be addressed to maintain the sustainability of these as well as their habitats. Moreover, with the advancement of new technologies such as the construction of water turbines for a small head of 2–3 m, it has now become possible to step up for a large number of small-scale plants, viz., micro and mini plants with capacities of up to 100 and 20,000 kW, respectively. This study can be useful to policymakers, environmentalists, technologists, and engineers in determining the factors to design and implement such programs.

References

Agarwal SS, Kansal ML (2020) Risk based initial cost assessment while planning a hydropower project. Energy Strategy Rev (31):100517

Ahlers R, Budds J, Joshi D, Merme V, Zwarteveen M (2014) Framing hydropower as green energy: Assessing drivers, risks and tensions in the Eastern Himalayas. Earth Syst Dyn Discuss 5:1521–1541

Alley KD, Hile R, Mitra C (2014) Visualizing hydropower across the Himalayas: mapping in a time of regulatory decline. HIMALAYA J Assoc Nepal Himalayan Stud 34(2):52–66

Altaf F, Meraj G, Romshoo SA (2013) Morphometric analysis to infer hydrological behaviour of Lidder watershed, Western Himalaya, India. Geogr J 2013

Arora M, Kumar R, Singh RD, Malhotra J, Kumar N (2016) Analysis of unusual meteorological conditions that led to recent floods in Bhagirathi Basin (Uttarakhand Himalayas). Hydrol Sci J 61(7):1238–1243

Aryal D, Wang L, Adhikari TR, Zhou J, Li X, Shrestha M, Wang Y, Chen D (2020) A model-based flood hazard mapping on the southern slope of Himalaya. Water 12(2):540

Balling RC, Keikhosravi Kiany MS, Sen Roy S, Khoshhal J (2016) Trends in extreme precipitation indices in Iran: 1951–2007. Adv Meteorol 2456809

Bell R, Fort M, Götz J, Bernsteiner H, Andermann C, Etzlstorfer J, Posch E, Gurung N, Gurung S (2021) Major geomorphic events and natural hazards during monsoonal precipitation 2018 in the Kali Gandaki Valley, Nepal Himalaya. Geomorphology 372:107451

Bhatt CM et al (2017) Satellite-based assessment of the catastrophic Jhelum floods of September 2014, Jammu & Kashmir, India. Geomat Nat Haz Risk 8(2):309–327

Bisht S, Sharma S (2018) Carbon footprints of Liquefied Petroleum Gas transportation in the Indian Himalaya. J Clean Prod 196:1065–1072

Bisht S, Sharma S, Chaudhry S (2016) Flash flood risk susceptibility in Gagas River Watershed-Kumaun Lesser Himalaya. Int J Adv Remote Sens GIS 5(5):1709–1725
Dewan TH (2015) Societal impacts and vulnerability to floods in Bangladesh and Nepal. Weather Clim Extremes 7:36–42
Dimri AP, Thayyen RJ, Kibler K, Stanton A, Jain SK, Tullos D, Singh VP (2016) A review of atmospheric and land surface processes with emphasis on flood generation in the Southern Himalayan rivers. Sci Total Environ
Dotti GR, Misra ON (1987) Himalaya energy system. Gyanodayaa Prakashan, Nainital, Uttarakhand, pp 70–82
Gujree I et al (2017) Evaluating the variability and trends in extreme climate events in the Kashmir Valley using PRECIS RCM simulations. Model Earth Syst Environ 3(4):1647–1662
Joy J, Kanga S, Singh SK (2019) Kerala flood 2018: flood mapping by participatory GIS approach, Meloor Panchayat. Int J Emerging Techn 10(1):197–205
Kanga S, Singh SK, Sudhanshu (2017) Delineation of urban built-up and change detection analysis using multi-temporal satellite images. Int J Recent Res Aspects 4(3):1–9
Kumar A, Kumar D (2008) Predicting direct runoff from hilly watershed using geomorphology and stream-order-law ratios: case study. J Hydrol Eng 13(7):570–576
Kumar A, Gupta AK, Bhambri R, Verma A, Tiwari SK, Asthana AKL (2018a) Assessment and review of hydrometeorological aspects for cloudburst and flash flood events in the third pole region (Indian Himalaya). Polar Sci 18:5–20
Kumar S, Singh SK, Mahendra RS (2018b) Assessment of land use and land cover during pre and post cyclone Phailin in Southern part of Odisha Coast. i-Manager's J Future Eng Technol 13(4):53
Kumar D, Katoch SS (2015) Sustainability assessment and ranking of run of the river (RoR) hydropower projects using analytical hierarchy process (AHP): a study from Western Himalayan region of India. J Mt Sci 12(5):1315–1333
Liu CW, Lin KH, Kuo YM (2003) Application of factor analysis in the assessment of groundwater quality in a blackfoot disease area in Taiwan. Sci Total Environ 313(1):77–89
Maikhuri RK, Nautiyal A, Jha NK, Rawat LS, Maletha A, Phondani PC, Bahuguna YM, Bhatt GC (2017) Socio-ecological vulnerability: assessment and coping strategy to environmental disaster in Kedarnath valley, Uttarakhand, Indian Himalayan Region. Int J Disaster Risk Reduction 25:111–124
Mazari RK, Sah MP (2005) Pulia nala cloudburst of July 16, district Kullu, Himachal Pradesh: lesson for policy implementation. Himalayan Geol 25(2):153–161
Meraj G et al (2018a) An integrated geoinformatics and hydrological modelling-based approach for effective flood management in the Jhelum Basin, NW Himalaya. Multidiscip Digital Publ Inst Proc 7(1):8
Meraj G et al (2018b) Geoinformatics based approach for estimating the sediment yield of the mountainous watersheds in Kashmir Himalaya, India. Geocarto Int 33(10):1114–1138
Meraj G et al (2015) Assessing the influence of watershed characteristics on the flood vulnerability of Jhelum basin in Kashmir Himalaya. Nat Hazards 77(1):153–175
Meraj G, Romshoo SA, Altaf S (2016) Inferring land surface processes from watershed characterization. Geostatistical and geospatial approaches for the characterization of natural resources in the environment. Springer, Cham, pp 741–744
Nandargi S, Dhar ON (2011) Extreme rainfall events over the Himalayas between 1871 and 2007. Hydrol Sci J 56(6):930–945
Nathawat MS et al (2010) Monitoring & analysis of wastelands and its dynamics using multiresolution and temporal satellite data in part of Indian state of Bihar. Int J Geomatics Geosci 1(3):297–307
Nepal S, Flügel WA, Shrestha AB (2014) Upstream-downstream linkages of hydrological processes in the Himalayan region. Ecol Process 3(1):1–16
Paul A, Khan ML (2010) Water: an eco-friendly and energy-efficient resource for milling in rural Arunachal Pradesh. Curr Sci 99(4):417

Prasad J, Prasad HS (2007) Performance evaluation of existing mini hydro power projects of Uttarakand—A case study. In: International conference on small hydropower-hydro Sri Lanka, vol 22, p 24

Prasad VH, Roy PS (2005) Estimation of snowmelt runoff in the Beas basin, India. Geocarto 20:2

Rather MA et al (2017) Assessing the influence of watershed characteristics on soil erosion susceptibility of Jhelum basin in Kashmir Himalayas. Arab J Geosci 10(3):59

Rawat KP, Pant CC, Tiwari PC, Pant PD, Sharma AK (2012) Spatial variability assessment of river-line floods and flash floods in Himalaya: a case study using GIS. Disaster Prev Manag 21(2):135–159

Sahu N, Sayama T, Saini A, Panda A, Takara K (2020) Understanding the hydropower and potential climate change impact on the himalayan river regimes—A study of local perceptions and responses from himachal Pradesh, India. Water 12(10):2739

Saini RP, and Kumar A (2006) Development of standard water mills in Uttaranchal. Himalayan Small Hydropower Summit, pp 267–278

Schwanghart W, Worni R, Huggel C, Stoffel M, Korup O (2016) Uncertainty in the Himalayan energy–water nexus: estimating regional exposure to glacial lake outburst floods. Environ Res Lett 11(7):074005

Sengupta DP (1987) Himalaya energy system. Gyanodayaa Prakashan, Nainital, India, pp 98–115

Sharma S, Kuniyal JC, Sharma JC (2007) Assessment of man-made and natural hazards in the surroundings of hydropower projects under construction in the beas valley of northwestern Himalaya. J Mt Sci 4(3):221–236

Shrestha MS, Takara K (2008) Impacts of floods in South Asia. J South Asia Disaster Study 1(1):85–106

Singh TP, Chandrashekhar J, Agrawal AK (2007) Analysis of water and sediment flow in desilting basin of a run-of-river hydroelectric project. In: International conference on small hydropower-hydro Sri Lanka, vol 22. p 24

Talchabhadel R, Panthi J, Sharma S, Ghimire GR, Baniya R, Dahal P, Baniya MB, KC S, Jha B, Kaini S, Dahal K, Kumar S (2021) Insights on the impacts of hydroclimatic extremes and anthropogenic activities on sediment yield of a river Basin. Earth 2(1):32–50

Weerakoon SB, Rathnayake U (2007) Effect of the entrance zone on the trapping efficiency of desilting tanks in run-of-river hydropower plants. In: International conference on small hydropower-hydro Sri Lanka, vol 22, p 24

Earthquakes in the Himalayas—Assessment and Forecasting

Synthetic Aperture Radar Interferometry to Measure Earthquake-Related Deformation: A Case Study from Nepal

Himanshu Verma, Yogendra Sharma, and Sumanta Pasari

Abstract The Himalayan syntax is located at the collision boundary of the Indian and the Eurasian plates form an arc that extends ~2400 km across the continent, starting from the Kashmir Himalaya to Northeast India. Due to this collision process, there have been several significant earthquakes in the Himalayan region. To monitor earthquake-related deformation, a space-based technology, known as interferometry synthetic aperture radar (InSAR), is often employed. In this study, we demonstrate the InSAR-derived deformation rates corresponding to a moderate earthquake that occurred in the western part of Nepal near Dipayal Silgadhi (29.323°N, 81.143°E) on 19 November 2019. We use 42 interferograms covering the epicentral area with latitude ranging from 29.1°N to 29.8°N and longitude ranging from 80.5°E to 81.8°E. We calculate the line of sight (LoS) velocity and the coherence for the area along with their residuals. As a preliminary result, we observe an average upliftment of ~17.6 ± 15.5 mm/yr and subsidence of ~13.2 ± 14.4 mm/yr along the study region.

Keywords InSAR · Earthquake · Interferogram · Radar · Nepal

1 Introduction

The Himalayan mountain range and surrounding Indo-Gangetic foreland basin formed as a result of the collision between the plates of India and Eurasia (Sharma et al. 2020). This collision is still continuing, and as a result, many earthquakes occur in the Himalayan region. In Nepal, there have been several earthquakes in the last two centuries such as the 1833 Kathmandu earthquake ($M_w = 7.3$), 1916 Nepal-Tibet earthquake ($M_w = 7.0$), 1934 Nepal-Bihar earthquake ($M_w = 8.0$) and the 2015 Gorkha earthquake ($M_w = 7.8$). The US Geological Survey (USGS) recorded a $M_w = 5.0$ earthquake around 21 km east-northeast of Dipayal Silgadhi on 19 November

H. Verma · Y. Sharma · S. Pasari (✉)
Department of Mathematics, Birla Institute of Technology and Science, Pilani, Pilani Campus, RJ 333031, India
e-mail: sumanta.pasari@pilani.bits-pilani.ac.in

S. Kanga et al. (eds.), *Disaster Management in the Complex Himalayan Terrains*, Geography of the Physical Environment,
https://doi.org/10.1007/978-3-030-89308-8_9

2019. The quake struck around 19:15 (local time) at a depth of 10 km. To understand the crustal deformation caused by this earthquake in the region, we used an InSAR-based approach.

The InSAR is a space-based technique that combines the phases of two different radar images obtained from satellites simultaneously or at different looking angles (Dumka et al. 2020; Yhokha et al. 2015; Kobayashi et al. 2015; Zhou et al. 2009; Morishita et al. 2020; Radutu et al. 2017; Gahalaut et al. 2017; Rajawat 2016; Ferretti et al. 2007). In this article, a fundamental concept of InSAR for earthquake-related deformation is described and presents a case study in Nepal that shows the unique potential of InSAR for earthquake-related deformation. We discuss a case study of InSAR in the field of earthquake monitoring. InSAR is also used in many other areas such as ocean currents, landscape surveys, vegetation parameters, classifications of terrain, glacial techniques and landslides (Massonnet and Feigl 1998).

To produce a radar interferogram, two or more SAR images can be combined. The interference pattern caused by the phase difference between these images can measure the topography or minute changes in the topography of the order of several millimetres between two image acquisitions along the satellite look path. A digital SAR image is a 2D account of the phases in the imaging region of the returns from targets. To generate an interferogram, a phase of two digital SAR images obtained simultaneously or at different viewing angles from space can be combined. Relative phase observations from two images captured from slightly distinct viewing angles provide information on changes in the range of targets on the area and can thus recover the topography at the pixel of the SAR image with information of the geometry of the imagery. The phase change between two SAR images captured from the same point of view, but at different times, is capable of reliably calculating any changes in the return phase. Therefore, if the Earth's crust changes between the two imaging passes towards or away from the radar, this would result in phase changes that can be calculated with an accuracy equal to displacements at the millimetre level.

The geometry of repeat-pass interferometry is provided by Fig. 1. Our objective is to identify, from the pair of SAR images, the elevation h of each pixel location. From the radar control system, the radar transmitter wavelength λ is identified. From appropriate satellite orbits, the satellite flying height H and orbit separation vector $\vec{B}$ can be calculated. To the radar signal delay, r_1 along the looking vector $\vec{l}_1$ is calculated and the interferogram is the relative change between the two-phase observations ϕ. The phase change calculated between the image pair is directly proportionate to the path delay change provided by $\phi = 4\pi/\lambda(r_1 - r_2)$, where $r_{1,2}$ is the $A_{1,2}$ antenna range. Therefore, if r_1 and $\vec{B}$ are known, the phase change can be determined. Only modulo 2π will evaluate this phase difference. Then, the interferometric phase is (Rosen et al. 2000)

$$\phi = \frac{4\pi r_1}{\lambda\left[1 - \frac{2\left(\vec{B}\cdot\hat{l}_1\right)}{r_1} + \left(\left(\frac{B}{r_1}\right)^2\right)^2 - 1\right]}$$

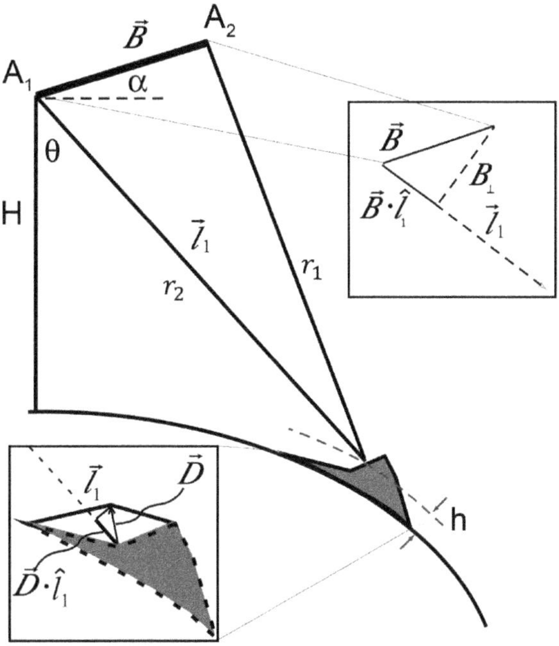

Fig. 1 Side-looking InSAR acquisition geometry (Bürgmann et al. 2000)

where $\hat{l}_1$ is the unit vector in the direction of the range and $\vec{B}$ is the vector of length B's baseline separation. A parallel ray approximation can be applied if $B \ll r_{1,2}$ then $\phi \approx -\frac{4\pi}{\lambda\left(\vec{B}\cdot\hat{l}_1\right)} = -\frac{4\pi}{\lambda B \sin(\theta-\alpha)}$ (Fig. 1) (Zebker and Goldstein 1986). We can calculate $\theta = \sin^{-1}\left(-\frac{\lambda\phi}{4\pi B}\right) + \alpha$ and $h = H - r_1 \cos\theta$, with the known wavelength of radar as demonstrate in Fig. 1.

It is useful to describe the uncertainty height h_a, the change in height that corresponds to a complete 2π shift in phase, to explain the impact of satellite orbital separation on the interferometric phase (Zebker and Goldstein 1986).

$$h_a = 2\pi\frac{\partial h}{\partial\phi} = \frac{\lambda r_1 \sin\theta}{2B\cos(\theta-\alpha)}$$

where $B\cos(\theta - \alpha) = B_\perp$ (Fig. 1).

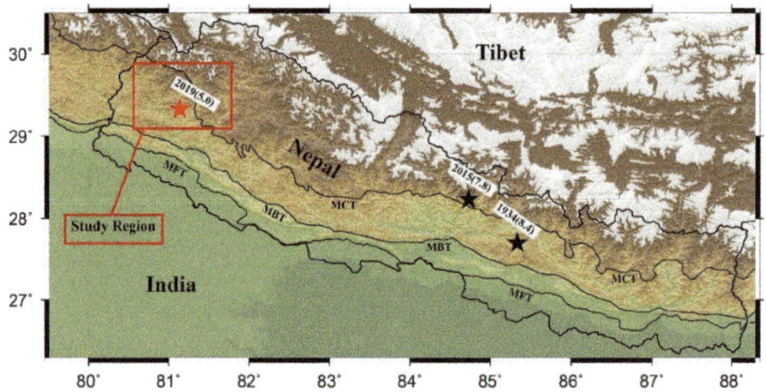

Fig. 2 Geology of Nepal and the red rectangle in the western part of Nepal highlights the present study area

2 Study Area

There are many evolving faults in Western Nepal such as the main boundary thrust (MBT), main central thrust (MCT), Himalayan frontal thrust (HFT) and the Mahabharat thrust (MT). The present study area falls near the MT and is bounded by 29.10°–29.80°N in latitude and 80.50°–81.80°E in longitude (Fig. 2). The study region comprises part of the higher Himalayan zone and the Tibetan-Tethys Himalayan zone.

3 Methodology

The deformation pattern based on the study of a set of SAR images has been monitored by differential interferometric synthetic aperture radar (DInSAR) method (Dumka et al. 2020; Yhokha et al. 2015; Kobayashi et al. 2015; Zhou et al. 2009; Morishita et al. 2020; Radutu et al. 2017; Gahalaut et al. 2017; Rajawat 2016; Ferretti et al. 2007). A flow chart of the proposed method is provided in Fig. 3 (Morishita et al. 2020). A master image has been elicited from the distribution of temporal baselines (Fig. 4) from various single look complex (SLC) images to maintain our dataset's coherence. A collection of sentinel SLC data (https://comet.nerc.ac.uk/COMET-LiCS-portal/) was processed for the interferogram generation for the present study using the LiCSBAS processing tool (Morishita et al. 2020). In order to estimate the milimetre level of velocity in the present study area, the interferometric phase change for these SAR images has been calculated.

We use the LiCSBAS processing tool to find the velocity (Fig. 5) and the time series of the correlated SAR data (as shown in the flow chart).

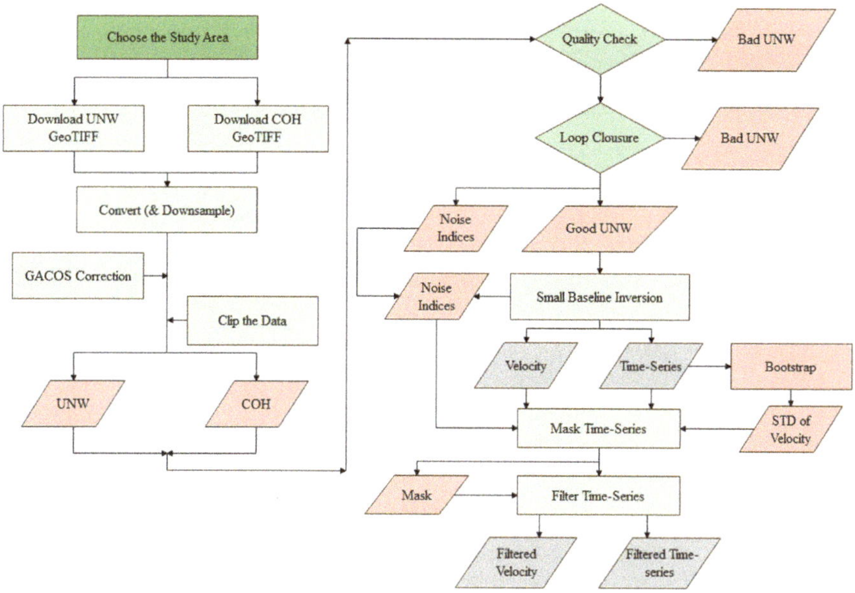

Fig. 3 Flow chart of the proposed methodology (Morishita et al. 2020)

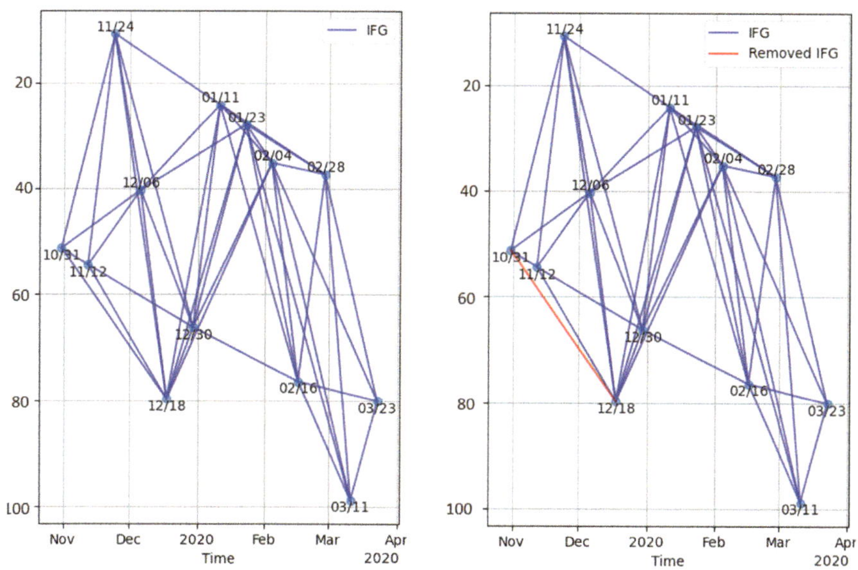

Fig. 4 The temporal baseline of SAR images. The red line shows removed uncorrelated interferogram (right figure)

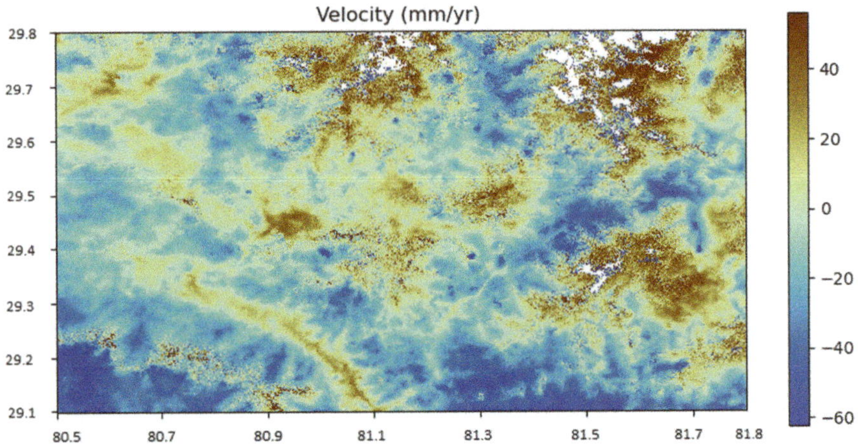

Fig. 5 The velocity of the present study region. The positive value and the negative value of velocity represent upliftment and subsidence, respectively

4 Results and Summary

The Himalayan subcontinent is one of the most seismically active zones in the world (Pasari 2015, 2018; Pasari and Sharma 2020). To understand the deformation pattern around the epicentral area of the Dipayal earthquake, the results are demonstrated from the proposed InSAR technique in this section. We use 42 interferograms from the study area from 31 October 2019 to 23 March 2020 to perform the DInSAR study. Around 9, 14, 000 pixels have been analysed for the measurement of velocity at different locations in the study area. The results reveal a mean upliftment of ~17.6 ± 15.5 mm/yr and a mean subsidence ~13.2 ± 14.4 mm/yr in the direction of the line of sight (LoS) (as shown in Fig. 5). After analysis of the velocity distribution, it can be observed that the location with latitude 29.20°N and longitude 80.52°E shows the maximum amount of subsidence, whereas, the location with 29.70°N latitude and 81.56°E longitude shows the maximum amount of upliftment. The time series of a location (29.35°N, 81.13°E), near the northeast of the earthquake epicentre, is shown in Fig. 6. Based on our findings, we infer that the MT fault may be considered as an active fault.

In summary, the present analysis leads to the following conclusions:

1. InSAR-based results in the present study mark the MT as an active fault.
2. With about five months of data set (31 October 2019 to 23 March 2020), the InSAR analysis comprising 9, 14, 000 pixels reveal a mean velocity of ~ 17.6 ± 15.5 mm/yr in terms of upliftment and ~13.2 ± 14.4 mm/yr in terms of subsidence in the direction of the line of sight (LoS).

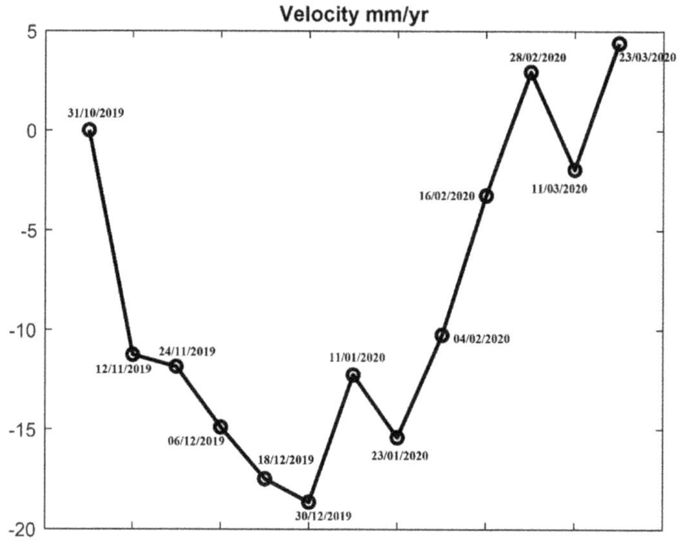

Fig. 6 The time series of around the location 29.35°N in latitude and 81.13°E in longitude

References

Bürgmann R, Rosen PA, Fielding EJ (2000) Synthetic aperture radar interferometry to measure Earth's surface topography and its deformation. Annu Rev Earth Planet Sci 28(1):169–209

Dumka RK, SuriBabu D, Malik K, Prajapati S, Narain P (2020) PS-InSAR derived deformation study in the Kachchh, Western India. Appl Comput Geosci 8:100041

Ferretti A, Monti-Guarnieri A, Prati C, Rocca F, Massonet D (2007) InSAR principles-guidelines for SAR interferometry processing and interpretation, TM-19. ESA Publications, The Netherlands

Gahalaut VK et al (2017) InSAR and GPS measurements of crustal deformation due to seasonal loading of Tehri reservoir in Garhwal Himalaya, India. Geophys J Int 209(1):425–433

Kobayashi T, Morishita Y, Yarai H (2015) Detailed crustal deformation and fault rupture of the 2015 Gorkha earthquake, Nepal, revealed from ScanSAR-based interferograms of ALOS-2. Earth Planets Space 67(1):1–13

Massonnet D, Feigl KL (1998) Radar interferometry and its application to changes in the Earth's surface. Rev Geophys 36(4):441–500

Morishita Y, Lazecky M, Wright TJ, Weiss JR, Elliott JR, Hooper A (2020) LiCSBAS: An Open-source InSAR time series analysis package integrated with the LiCSAR automated sentinel-1 InSAR processor. Remote Sens 12(3):424

Pasari S (2015) Understanding Himalayan tectonics from geodetic and stochastic modeling. Ph.D. thesis, Indian Institute of Technology Kanpur, India, pp 1–376 (2015)

Pasari S (2018) Stochastic modeling of earthquake interoccurrence times in northwest Himalaya and adjoining regions. Geomat Nat Haz Risk 9(1):568–588

Pasari S, Sharma Y (2020) Contemporary earthquake hazards in the west-northwest Himalaya: a statistical perspective through natural times. Seismol Res Lett 91(6):3358–3369

Radutu A, Nedelcu I, Gogu CR (2017) An overview of ground surface displacements generated by groundwater dynamics, revealed by InSAR techniques. Procedia Eng 209:119–126

Rajawat A (2016) Crustal deformation studies using synthetic aperture interferometry. Proc Ind Natl Sci Acad 82(3):737–746

Rosen PA, Hensley S, Joughin IR, Li FK, Madsen SN, Rodriguez E, Goldstein RM (2000) Synthetic aperture radar interferometry. Proc IEEE 88(3):333–382

Sharma Y, Pasari S, Ching KE, Dikshit O, Kato T, Malik JN, Chang CP, Yen JY (2020) Spatial distribution of earthquake potential along the Himalayan arc. Tectonophysics 791:228556

Yhokha A, Chang CP, Goswami PK, Yen JY, Lee SI (2015) Surface deformation in the Himalaya and adjoining piedmont zone of the Ganga Plain, Uttarakhand, India: Determined by different radar interferometric techniques. J Asian Earth Sci 106:119–129

Zebker HA, Goldstein RM (1986) Topographic mapping from interferometric synthetic aperture radar observations. J Geophys Res: Solid Earth 91(B5):4993–4999

Zhou X, Chang NB, Li S (2009) Applications of SAR interferometry in earth and environmental science research. Sensors 9(3):1876–1912

Earthquake Forecasting in the Himalayas Artificial Neural Networks

Arnav Ahuja and Sumanta Pasari

Abstract Earthquake is a natural phenomenon that causes huge loss in both life and property. Improvement of seismic hazard assessment requires integrated techniques such as geodetic, stochastic, and machine learning models. Forecasting of the time of the event, magnitude, and location of the epicenter of future events has been the major focus of several efforts in recent years. Many methods have been proposed to forecast the occurrence of earthquakes like statistical methods and other modeling approaches. Such methods are based on either the study of electric or magnetic signals or microseismicity patterns in which changes are experienced due to an upcoming event. In this study, our aim is to forecast earthquakes using neural networks, based on some seismicity indicators which capture the intrinsic information of the earthquake events. For this, an effective neural network architecture is created with different deep learning optimization algorithms and the results showed that the eight seismicity indicators have essentially captured most of the information of earthquake events. It is observed that neural networks are an effective tool for forecasting earthquakes as the neural networks well capture the nonlinearity and heterogeneity of inherent mechanisms with appropriate weights. The proposed network provides 90% accuracy and an F1-score of 0.89. It is hoped that this study shall provide useful information to the industry, academia, and government agencies to develop new standards of monitoring and mitigation measures of earthquake hazard.

Keywords Earthquakes · Himalayas · Risk · Vulnerability · Seismicity · Artificial neural networks

A. Ahuja · S. Pasari (✉)
Department of Mathematics, Birla Institute of Technology and Science, Pilani, Pilani Campus, RJ 333031, India
e-mail: sumanta.pasari@pilani.bits-pilani.ac.in

S. Kanga et al. (eds.), *Disaster Management in the Complex Himalayan Terrains*, Geography of the Physical Environment,
https://doi.org/10.1007/978-3-030-89308-8_10

1 Introduction

Earthquake is a natural phenomenon that causes huge loss in both life and property. Since ancient times, we have always been looking for ways to prevent or control these natural disasters. In the northern part of India, the entire Himalayan subcontinent is prone to earthquakes due to the persistent collision between the Indian plate and the Eurasian plate at an average rate of 4–5 cm/yr (Sharma et al. 2020, 2018). Improvement of seismic hazard assessment requires integrated techniques such as geodetic, stochastic, and machine learning models.

Occurrence of earthquakes is very common in the Himalayan regions. In about a decade, the Himalayan orogen has so far experienced five disastrous earthquakes, namely the April 4, 1905 Kangra earthquake (Mw 7.8), the January 15, 1934 Bihar-Nepal earthquake (Mw 8.1), the August 15, 1950 Assam earthquake (Mw 8.4), October 8, 2005 Kashmir earthquake (Mw 7.6), and the most recent April 25, 2015 Gorkha, Nepal event (Mw 7.8) (Sharma et al. 2020, 2018; Pasari 2015, 2018). While the Gorkha earthquake in Nepal caused a staggering death toll of about 9000 people and destroyed around 600,000 houses, the Kashmir event in 2005 caused a death toll of around 80,000 with the collapse of around 32,000 buildings. These two events in contemporary times are dramatic reminders of Himalayan seismicity. Moreover, the geodetic studies reveal that one or more great earthquakes are due in the Himalayan orogen (Sharma et al. 2020, 2018). Therefore, earthquake forecasting is one of the most important tasks for the safety of millions of people in the Himalayan foothills and adjoining regions.

Forecasting of the time of the event, magnitude, and location of the epicenter of future events has been the major focus of several efforts in recent years (Sharma et al. 2020, 2018; Pasari 2015, 2018). Many methods have been proposed to forecast the occurrence of earthquakes like statistical methods and other modeling approaches. Such methods are based on either the study of electric or magnetic signals or micro-seismicity patterns in which changes are experienced due to an upcoming event (Scholz 2019). Historical data from seismic catalogs play a dominant role to help develop mathematical tools to forecast these events.

The exponentially rising volume of seismic data in recent years has provided us with the opportunity to involve deep learning algorithms to detect and locate earthquakes reliably. By the universal approximation theorem, neural networks can model any complex function (Asim et al. 2017; Azam et al. 2014). Therefore, such methods must be able to infer the complex relationship between the earthquake processes and the occurrence phenomenon of earthquakes. Neural networks perform a key task related to the earthquake prediction in terms of its magnitude, time, or location. In this study, our aim is to forecast earthquakes using neural networks, based on some seismicity indicators (e.g., (Panakkat and Adeli 2007)) which capture the intrinsic information of the earthquake events. For this, an effective neural network architecture is created with different deep learning optimization algorithms and the results are discussed.

2 Model Formulation

This section provides basic formulation steps of machine learning algorithms for seismicity forecasting. It also mentions the seismic indicators along with their usefulness.

2.1 *Neural Networks*

Comprising a chain of algorithms, the neural networks imitate the function of a human brain to identify the inherent associations between large volumes of dataset. These algorithms are essentially a series of transformations to the input to get the output with the transformations depending on some parameter which are found out by learning from the data during a process called training. The network then adjusts its parameters according to a learning rule and using the error value from the cost function defined later. Successive maps will result in producing results that are increasingly similar to the real output. After sufficient number of these maps, the process of training can be terminated upon reaching a certain criterion. The process is known as supervised learning. These techniques prove their usefulness in all streams of science and engineering, ranging from financial forecasting, marketing research of fraud detection, to risk analysis for a multi-strategic planning. Some important terminologies related to neural network are mentioned below.

Hyperparameters Hyperparameters are constant parameters whose values are set before the training process begins. Examples of hyperparameters include the number of layers, learning rate, and batch size.

Backpropagation It is a method to change the values of parameters of the network during the training process. It controls how much the value of the learning rate will be determined by the hyperparameter learning rate as well as the cost function.

Cost Function This is a measure of how different the prediction is from the actual value of the output. There can be different types of cost functions such as the mean squared loss and the cross entropy loss.

Training Training the network from a given data point is usually performed by finding the difference between the output of the neural network (often a prediction) and a target output. This is the cost computed from the cost function. After computing the cost, backpropagation is applied to change the parameters of the networks and get a new prediction from the network. The equation is as follows:

$$w_{i+1} = w_i - l_r \times \Delta w$$

Here, w_{i+1} is the new weight parameter, l_r is the learning rate of hyperparameter, and Δw is the derivative of the cost function with respect to the weight parameter.

Epoch One pass over the entire dataset during the training process is called epoch. Usually, the network is trained over multiple epochs to minimize the cost function.

Adam's Algorithm for Backpropagation This is an optimization algorithm for the learning rate to effectively train deep networks. The algorithm uses adaptive learning rates method to determine separate learning rate of each parameter. The formula for the Adam's algorithm is

$$\begin{aligned} v_t &= \beta_1 v_{t-1} - (1-\beta_1) g_t \\ s_t &= \beta_2 s_{t-1} - (1-\beta_2) g_t^2 \\ \Delta w_t &= -\eta v_t \big/ (s_t + \varepsilon)^{0.5} \\ w_{t+1} &= w_t + \Delta w_t \end{aligned}$$

η Initial learning rate;
g_t Gradient at time t along the parameter;
v_t Exponential average of gradients along the parameter;
s_t Exponential average of squares of gradients along the parameter;
$\beta_1,\ \beta_2$ Hyperparameters.

2.2 *Seismicity Indicators*

We use eight seismicity indicators which are calculated from the time series data (e.g., (Panakkat and Adeli 2007)) to form the input of the proposed neural network. The intuition behind the calculation of these seismicity indicators comes from the Gutenberg Richter (G-R) empirical law. A brief description of the seismicity indicators is provided below.

Among the three basic empirical laws in seismology, namely the G-R law, Omori's aftershock decay, and the Bath's law, G-R law addresses the frequency-magnitude power law distribution in a seismogenic area during a fixed time domain. According to this law, the number of earthquakes are exponentially distributed to the magnitude of events. In other words, for constants a and b, the G-R equation can be written as

$$\log_{10} N(M) = a - bM$$

where $N(M)$ denotes the number of events of magnitude larger than or equal to M, a is called the "productivity" of the region, and b is the slope of the line, known as the "b-value," typically lies in the range of 0.8–1.1 (Scholz 2019; Pasari and Sharma 2020; Pasari 2019a; Bhatia et al. 2018; Pasari and Mehta 2018).

For the present analysis, we consider eight seismicity indicators (Panakkat and Adeli 2007) as input to our network as they capture a large amount of the seismic information of the earthquake event. Brief description of these indicators is mentioned below.

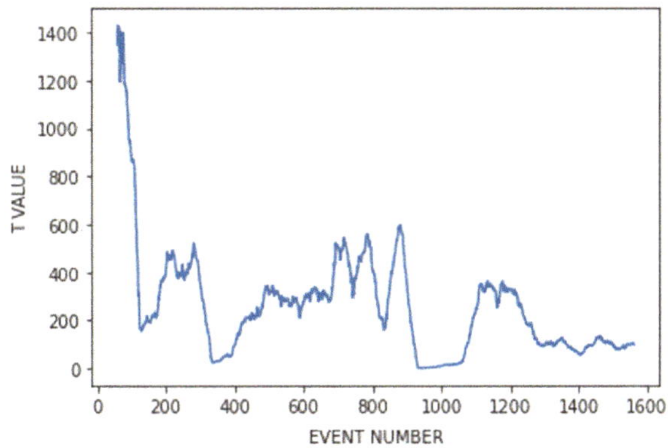

Fig. 1 A plot of T-value versus event number

The T-value The time elapsed for events greater than a predefined magnitude threshold among the last n events is defined as

$$T = t_n - t_1,$$

Here, t_n and t_1 denote the occurrence times for the nth and 1st events, respectively. A large T-value indicates that there are a smaller number of foreshocks which further means that there is a lower probability of an upcoming large earthquake. Conversely, a smaller T-value implies that the frequency of the foreshocks is high which signals toward a high probability of an upcoming large earthquake. Figure 1 presents the concept.

The Mean Magnitude The average of magnitudes for the last n events is given by

$$M_{\text{mean}} = \frac{1}{n}\sum_{i=1}^{n} M_i$$

As the foreshock magnitudes increase prior to a large earthquake, a high mean magnitude denotes increased probability of a forthcoming seismic event. Figure 2 illustrates the concept.

The Rate of Square Root of Released Seismic Energy If E amount of seismic energy gets released over a time T, then the following relations are defined.

$$dE^{1/2} = \sum E^{1/2} \Big/ T \qquad \text{and} \qquad E = 10^{(11.8+1.5M)}\text{ergs}$$

The above relations are reasonable as the geological fault systems store energy that is released suddenly in the form of a major earthquake event after reaching to the bearing strength of a fault. A higher energy release suggests a forthcoming event. Figure 3 shows the concept.

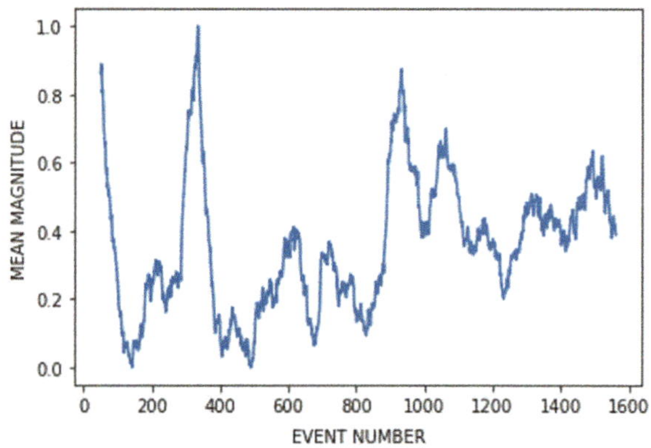

Fig. 2 A plot of mean magnitude and event number

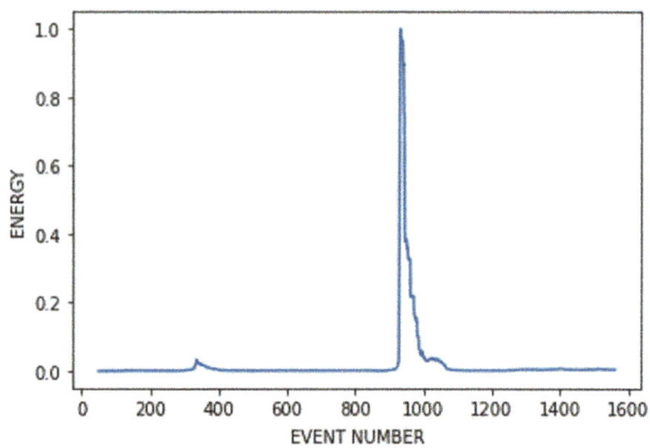

Fig. 3 A plot of seismic energy and number of events

G-R b-value The b-value in the G-R relation, $\log_{10} N(M) = a - bM$, serves as an important parameter in defining the characteristics of the seismic event.

Sum of the Mean Square Deviations from the Regression Line Based on the G-R Inverse Power Law (η Value) The η-value is defined as

$$\eta = \frac{1}{n-1} \sum_{i=1}^{n} \left(\log_{10} N_i - (a - bM_i)\right)^2$$

This measures the deviation of the original data from the G-R empirical relation. A lower value of η means that the original data is close to the G-R relation, whereas, a higher value suggests randomness. Thus, a higher value indicates inability of the G-R empirical law to favorably model the underlying distribution (Fig. 4).

Difference Between the Largest Observed Magnitude and Expected Magnitude (Magnitude Deficit) Based on G-R Law (ΔM Value) The ΔM value can be expressed as

Fig. 4 A plot of η value versus the number of events

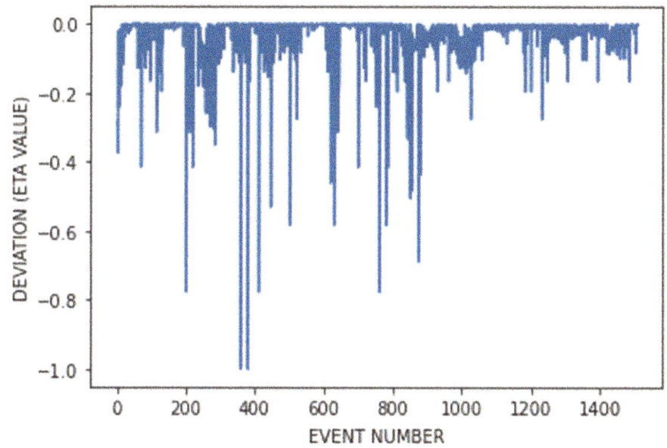

$$\Delta M = M_{\text{max,obs}} - M_{\text{max,exp}}$$

Here, $M_{\text{max,obs}}$ denotes the maximum observed magnitude (in the last n events), whereas, $M_{\text{max,exp}}$ denotes the expected maximum magnitude from G-R law. Notice that in a catalog, as the maximum magnitude usually occurs once, we put $N = 1$ in the G-R relation. This yields $M_{\text{max,exp}} = a/b$.

Mean Earthquake Interevent Time (μ Value) As characteristic events ideally differ by approximately equal time periods (e.g., (Scholz 2019)), the mean interevent time (μ) for the n large events can be expressed as

$$\mu = \sum_{i=1}^{n} t_i / n$$

Coefficient of Variation of Mean Interevent Times (μ), Called as the Aperiodicity of the Mean (c Value) To determine the closeness of the distribution of observed magnitudes from that of the characteristic events, the c value is defined as.

$$c = \text{standard deviation of the observed times}/\mu$$

A larger value of the aperiodicity shows a higher difference between the calculated mean time and the observed mean time, and vice-versa.

2.3 *Design of a Suitable Network*

The neural networks are often found out after several hit and trial runs along with a systematic reasoning, as mentioned below.

If the training accuracy turns out to be high but the validation accuracy is low, then it means that the model is over fitting, indicating that the model so comprehensive that it is learning the training data extremely well. Hence, it is over fitting to the training data though giving low accuracy on the validation data. In such a case, decreasing the number of layers is the right approach. Regularization can also be used here.

If the training accuracy as well as the validation accuracy are low, then it means that the model is under fitting which suggests that the model is not complex enough to capture the original relationship between the input and output. In such a case, increasing the complexity of the model by increasing the number of layers could work.

If the size of the dataset is very big, then sometimes the total number of features can be less as compared to the size of the dataset. For example, use only 2 input features and train the network with 30,000–40,000 points. This can be adapted by splitting the data into random batches and then training them separately and ultimately assembling them; otherwise, train a smaller subset of the dataset of random points (but we have to be careful while choosing the subset) once and then test it on random batches of data.

3 Data and Implementation

The earthquake data of Indian Himalaya from the year 1980 to 2020 are collected from the International Seismological Center (ISC) bulletin. The catalog consists of 1562 events ($M \geq 3$) with the information of occurrence date/time, latitude, longitude, depth, and magnitude. Values of the above mentioned eight indicators are then calculated for the present dataset. These inputs are normalized using the min–max scaling to achieve faster convergence of the model.

The proposed neural network comprises 8 neurons in input layer, 24 in first hidden layer, 24 in second hidden layer, and 12 in third hidden layer. The activation functions are decided accordingly. It is trained using 1132 data points (i.e., earthquake events). The training process involves getting the loss value and the accuracy of the prediction from the model. Finally, the loss value of the prediction model is used to backpropagate the network and change the weights of the network according to the Adam optimizer. The loss is calculated through the binary cross entropy function, as there are only two classes. The backpropagation algorithm used is Adam's algorithm along with mini-batch technique to get the maximum accuracy. The mini-batch size is 64 and it is determined empirically to obtain the maximum accuracy. The binary cross entropy loss function defines a loss value of the output class when there are two output classes, as is the case here. The code for training the networks is available upon request to the authors. The output dataset consists of 1 for events which have an event of magnitude 4.3 or greater in the next 30 days, otherwise it returns 0.

4 Results

Results of the proposed neural network are discussed in this section. We use accuracy, F1-score, and learning curve of the model to illustrate the results.

4.1 Accuracy

Accuracy is one of the several desirable metrics to evaluate the performance of the classification model. It is nothing but the ratio of the correct predictions and the total number of predictions. The proposed model has got about 90% accuracy.

4.2 F1-Score

While the accuracy is used when the distribution of classes is similar, the F1-score provides a better realization for an imbalanced distribution of classes as in this study. In the case of earthquakes, the dataset is often imbalanced as the frequency of high magnitude earthquakes is less and frequency of low magnitude earthquakes is high. The F1-score, the conventional or the balanced F-score, is the harmonic mean of the precision as well as recall and is defined as

$$\mathrm{F1} = \frac{\mathrm{TP}}{\mathrm{TP} + 0.5\,(\mathrm{FP} + \mathrm{FN})}$$

Here, TP, FP, and FN denote the true positives, false positives, and the false negatives. The F1-score achieved of the Himalayan region is 0.89.

4.3 Learning Cure for the Model

The models were trained using the Adam's optimizer as it gives a faster and better convergence than the traditional stochastic gradient descent. Figure 5 shows the loss curves for the Himalayan region.

5 The Way Forward and Conclusions

Earthquakes are inevitable phenomena in the Himalayan subcontinent, resulting in huge damages to life and property (Sharma et al. 2018, 2020; Pasari 2015, 2018).

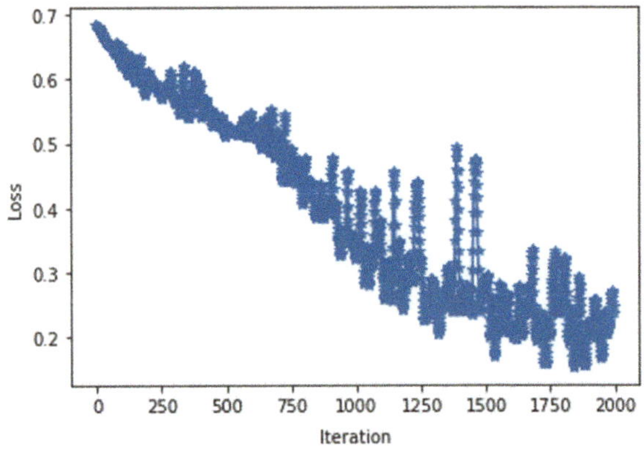

Fig. 5 Loss curve for the study region

To mitigate earthquake risks, a number of methods have been proposed (Sharma et al. 2020, 2018; Pasari 2015, 2018, 2019a, b; Scholz 2019; Asim et al. 2017; Azam et al. 2014; Panakkat and Adeli 2007; Pasari and Sharma 2020; Bhatia et al. 2018; Pasari and Mehta 2018; Pasari and Dikshit 2014a, b, 2015a, b, 2018; Petersen et al. 2007). The prediction of earthquakes in the Himalayan region has been addressed in this paper by the means of neural networks. The G-R law is a useful tool to estimate the relationship between the number of earthquakes and their magnitudes. The eight seismicity indicators have essentially captured most of the information of earthquake events. It is observed that neural networks are an effective tool for forecasting earthquakes as the neural networks well capture the nonlinearity and heterogeneity of inherent mechanisms with appropriate weights. The proposed network provides 90% accuracy and an F1-score of 0.89.

The proposed methodology and the emanated results provide useful information to the industry, academia, and government agencies to develop new standards of monitoring and mitigation measures of earthquake hazard (Pasari and Sharma 2020). It leads to better seismic hazard management as well as planning and design considerations of earthquake insurance, large scale construction planning, or public utilities in the Himalayan foothills. Satellite remote sensing, geographical information system (GIS), and other techniques may be combined with the proposed method for improved pre- and post-management of earthquakes. In addition, awareness campaigns must be carried out among the locals in the seismically active Himalayan belt. Common survival measures like storing food and water for emergency, and knowledge of what to do during and after an event are essential. Each individual must strictly follow the seismic design codes while constructing houses, public buildings, and bridges, along with modifying and repairing older structures for increased resistance (Pasari 2015; Scholz 2019).

The government also needs to allocate more funds to the research of these natural disasters, so as to efficiently combine several techniques for a fruitful result. Along with the understanding of earthquake kinematics using enhanced space geodetic methods, stochastic and machine learning based empirical methods must be encouraged for earthquake pattern analysis. Methods like satellite remote sensing and GIS are also efficient ways to provide a framework for gathering and analyzing data for multistate decision making and actionable planning in case of an upcoming disaster.

Though the proposed study through neural network provides satisfactory outputs, a high imbalance is evident in the datasets as earthquakes of higher magnitudes do not occur as frequently as those of lower magnitudes. In this sense, the future work may address data augmentation techniques to generate artificial events to make the model more robust to higher magnitude earthquakes.

References

Asim KM, Martínez-Álvarez F, Basit A, Iqbal T (2017) Earthquake magnitude prediction in Hindukush region using machine learning techniques. Nat Hazards 85(1):471–486

Azam F, Sharif M, Yasmin M, Mohsin S (2014) Artificial intelligence based techniques for earthquake prediction: a review. Sci Int 26(4):1495–1502

Bhatia A, Pasari S, Mehta A (2018) Earthquake forecasting using artificial neural networks. ISPRS-Int Arc Photogram Rem Sensing Appl 42(5):823–827

Panakkat A, Adeli H (2007) Neural network models for earthquake magnitude prediction using multiple seismicity indicators. Int J Neural Systems 17(1):13–33

Pasari S, Sharma Y (2020) Contemporary earthquake hazards in the west-northwest Himalaya: a statistical perspective through natural times. Seismol Res Lett 91(6):3358–3369

Pasari S (2019a) Nowcasting earthquakes in the Bay of Bengal region. Pure Applied Geophysics 176(4):1417–1432

Pasari S (2019b) Inverse Gaussian versus lognormal distribution in earthquake forecasting: Keys and clues. J Seism 23(3):537–559

Pasari S, Mehta A (2018) Nowcasting earthquakes in the northwest Himalaya and surrounding regions. ISPRS-Int Arc Photogram Rem Sensing Appl 42(5):855–859

Pasari S, Dikshit O (2014a) Three-parameter generalized exponential distribution in earthquake recurrence interval estimation. Nat Hazards 73:639–656

Pasari S, Dikshit O (2014b) Impact of three-parameter Weibull models in probabilistic assessment of earthquake hazards. Pure Appl Geophys 171(7):1251–1281

Pasari S (2015) Understanding Himalayan tectonics from geodetic and stochastic modeling. Ph.D. thesis, Indian Institute of Technology Kanpur, India, pp 1–376

Pasari S (2018) Stochastic modeling of earthquake interoccurrence times in northwest Himalaya and adjoining regions. Geomat Nat Haz Risk 9(1):568–588

Pasari S, Dikshit O (2015a) Distribution of earthquake interevent times in northeast India and adjoining regions. Pure Appl Geophys 172(10):2533–2544

Pasari S, Dikshit O (2015b) Earthquake interevent time distribution in Kachchh, northwestern India. Earth Planets Space 67:129

Pasari S, Dikshit O (2018) Stochastic earthquake interevent time modelling from exponentiated Weibull distributions. Nat Hazards 90(2):823–842

Petersen, M. D., Cao, T., Campbell, K. W., & Frankel, A. D.: Time-independent and time-dependent seismic hazard assessment for the state of California: Uniform California earthquake rupture forecast model 1.0. Seismol Res Lett 78(1), 99–109 (2007).

Scholz CH (2019) The mechanics of earthquakes and faulting. Cambridge University, Cambridge
Sharma Y, Pasari S, Ching KE, Dikshit O, Kato T, Malik JN, Chang CP, Yen JY (2020) Spatial distribution of earthquake potential along the Himalayan arc. Tectonophysics 791:228556
Sharma Y, Pasari S, Dikshit O, Ching KE (2018) GPS-based monitoring of crustal deformation in Garhwal-Kumaun Himalaya. ISPRS-Int Arc Photogram Rem Sensing Appl 42(5):451–454

Hazard Mitigation Strategies in the Himalayas

Forest Fire Alert System of India with a Special Reference to Fire Vulnerability Assessment of the UT of Jammu and Kashmir

Majid Farooq, Soheib Gazali, Mudasir Dada, Neelu Gera, and Gowhar Meraj

Abstract This chapter discusses the forest fire alert system of India with a special reference to fire vulnerability assessment of the UT of Jammu and Kashmir. Using actual forest fire incidences from the Jammu and Kashmir state Forest Department for the period, 2002–2018 and correlating it with the MODIS satellite fire data (2012–2018) a comprehensive forest fire vulnerability assessment of the UT of Jammu and Kashmir has been carried out. Correlation between MODIS based (3141), and actual fire incidences (2438) from 2012 to 2018 showed a correlation coefficient of 0.97 even using a seven-year dataset of the actual fire points. Although, the points derived from MODIS satellite data cover a larger spatial extent the high correlation coefficient statistically provides a proof for using MODIS-based fire incidence data as a parameter for evaluating zones under different categories of forest fire vulnerability in Jammu and Kashmir. The results showed many regions of the UT fall in high fire vulnerability category and can only be managed if forest fire alert system of the UT is followed very strictly.

Keywords Forest fire · MODIS · Kashmir Himalaya · Vulnerability · Jammu and Kashmir

1 Introduction

Forests are an integral part of people's lives. They are humanity's most unbelievable wealth of nature and play an essential role in its existence. They have been a substantial source of food, wood, and many other products, as well as providing shelter and protection to many living organisms (Meraj et al. 2015, 2016). Fire is an integral part of the operation of many forest ecosystems and one of the major natural forces that have affected the populations of plants over time. For thousands of years, people have been using fire as a land management tool. The health of many habitats as a natural process plays an essential role. Changes in humans' fire dynamics and the

M. Farooq · S. Gazali · M. Dada · N. Gera · G. Meraj (✉)
Department of Ecology, Environment and Remote Sensing, Government of Jammu and Kashmir, Srinagar 190018, India

S. Kanga et al. (eds.), *Disaster Management in the Complex Himalayan Terrains*, Geography of the Physical Environment,
https://doi.org/10.1007/978-3-030-89308-8_11

increase in the rate of extremely high temperatures have nevertheless contributed to a situation of great threat to many forests and their biodiversity in the latter part of the twentieth century (Kanga et al. 2017a, b; Joy et al. 2019). Fire is also adversely affected and directly destroys forest trees, regenerating forests, microclimates, soil erosion, animal species, etc. Forest fire causes forest vegetation to return in most situations. Forest fires are a hazard to the richness of forests and the whole regime for fauna and flora, which seriously affect the region's biodiversity, ecology, and climate (Bhatt et al. 2017; Gujree et al. 2017). A forest fire can be described as non-contained combustion, which eats and freely spreads natural fuels (Rather et al. 2017).

In the Jammu region's subtropical forests, forest fires are common, while in the Kashmir region, they are low in intensity and frequency. However, the environment that leads to forest fires and global climatic change, such as low, irregular winter plasters, dry summer and spring seasons, create. Forest fires cause considerable environmental, economic, and social problems in many countries and potentially have long-term impacts on the natural environment and the economy.

India is one of the world's largest biodiversity areas, full of unique, diverse flora and fauna (Kanga et al. 2017; Kumar et al. 2018). The total forest area of the country is 764,566 km^2. Almost 40,000 acres of forest fire was struck in the Bandipur Tiger Reserve and National Park in February 2019 to kill the forest guard and hurt many others, particularly those who tried to exterminate the fire. It had almost engulfed most northern and western parts of the reserve forest. There are numerous forest fire incidents reported every year throughout the country. In 2003–2016, 380 and 445 in the 647 districts of the country had fires each year between 2003 and 2016 (i.e., at least 59%, although not less than 69%) according to the Moderate Resolution Imaging Spectroradiometer (MODIS) satellite detection of forest fires (MoEFCC 2018).

According to the Intergovernmental Panel on Climate Change's Fifth Assessment Report, carbon emissions from forest fires are from 2.5 billion to 4.0 billion tons of CO_2 per year (MoEFCC 2018). Wild forest fires have a global impact on wildlife habitats and biodiversity (Rather et al. 2017). Those fires also displace millions of people and take livelihoods away. A forest fire can be divided into three stages: grass, subterranean, and firefighting. The grass and the dry soil typically first ignite and are easy to handle. If the fire reaches the treetops, mainly at conifers, resulting in a crown or canopy fire, it is more challenging to check. Only about 4% of the world's forest fires, such as lightning, have natural causes. The anthropogenic origin of fires is common in all other cases. The burned areas are usually lost irrevocably and the whole ecosystem with its plants and animals. Depending on the type of plant, climate, and various other factors, the fire season varies from place to place. Even if the country's main forest fire season runs from February to June, some forests are not fire-free all year round. According to the India State of Forest Report (ISFR) 2015, the estimate for heavy, moderate, and mild fire exposure areas is 2.40, 7.49, and 54.40%, which makes for 64.29% of the total forest fire exposure area recorded.

2 Use of GIS in Forest Fires Mapping

Forests or wildfires are considered severe problems that distress many terrestrial ecosystems and cause ecological and economic damage, such as lack of land use revenue, destruction and property loss, damage to agriculture, and biodiversity loss. It is also one of the most significant causes of the depletion of the soil (Altaf et al. 2013).

For the effective and sound decision-making process of forest management, information on the distribution of forest fire risk zones is essential. A critical part of fire prevention is forest fire risk assessment. Pre-fire planning tools require objective instruments to track when and where a fire is more likely to occur or when it will have more adverse effects. Modeling of any land system process forest fires includes risk assessment and evaluation (Kanga et al. 2020). The word "risk" is used to describe the possibility that a fire will start as influenced by the presence and occurrence of causative agents (Altaf et al. 2014).

In India, fire affects 2–3% of the forest area annually (Roy 2003). On average, per year, more than 34,000 ha of forested areas are destroyed by fire. Since wildfires rely on spatial parameters, efficient use of remote sensing and geographic information systems may help plan risk maps for wildfires based on particular requirements and parameters and establish suitable solutions.

Remote sensing has been widely applied to obtain them due to the lack of accurate mapping of fire boundaries occurring in many areas. Once this data is accomplished, fire regimes and plant communities' regeneration are analyzed in many world regions. Remote sensing by satellite has become a primary source of data for predicting fire danger ratings, fuel and fire mapping, fire monitoring, and fire ecology. Satellite remote sensing techniques, as cost-effective instruments, play an essential role in fire mapping (Rather et al. 2017). Improved remote sensing techniques can update older fire scars and provide burn severity estimates. Remote sensing by satellite is well suited for assessing the extent of biomass burning, a prerequisite for estimating regional and global emissions needed better to understand the effects of fire on climate change (Meraj et al. 2018a, b).

3 National Policies on Forests and Forest Fires

3.1 National Forest Policy

The 1988 National Forest Policy seeks to put 33% of the country's geographical area under forest and tree cover (Dhyani and Handa 2013). The National Forest Policy also outlines the forest fire occurrence strategy and reports that it is high in the region. On a large scale, standing trees and fodder are burned, and such fires annihilate natural regeneration. During the fire season, special precautions must be taken. Improved management practices should be implemented to cope with forest fires.

3.2 National Action Plan on Forest Fires, 2018

Recognizing the need to revamp the country's forest fire management, the Government of India's Ministry of Environment, Forest and Climate Change (MoEFCC) has created the 2018 National Action Plan on Forest Fires. The action plan's key goals are to minimize forest fires by educating, allowing, inspiring, and promoting forest fringe communities to work with the State Forest Departments (SFDs) in tandem. The plan also aims to reduce the vulnerability of forests to fire hazards across the country's diverse forest ecosystems, strengthen institutions' firefighting capabilities, and speed up recovery after a fire incident.

By educating, facilitating, and empowering forest fringe communities and encouraging them to work in partnership with the Forest Departments, the National Action Plan on Forest Fires envisages minimizing forest fires. This will significantly reduce forests' vulnerabilities across the diverse forest ecosystems in the Indian subcontinent against fire hazards, boost forests and other workers' and institutions' capacity to battle fires, and speed up recovery after a fire case. NAPFF emphasizes the theoretical rationale for defining priority management intervention areas, allocating resources to priority areas, and tracking fire risk mitigation or control steps' efficacy.

3.3 National Forest Fire Alert System

Spatial data (MODIS) is used to locate and monitor forest fires in the nascent stage and provide SFDs and the general public with rapid and accurate signals to initiate preventive measures at their end (Fig. 1). In 2016, the Forest Survey of India established an indigenous Forest Fire Early Warning Alert System. State Forest Department warnings focus on forest cover, form of forest, environment variables (temperature and rainfall), and recent area fire incidences. It overlays the GIS layers of these parameters. Intersect regions above threshold values are selected and transmitted via email to the State Forest Departments' nodal officer as pre-warning forest fire warnings in the form of KML files. For the subsequent week, these warnings produced based on short-term weather variables are accurate. In 2017, this procedure was further refined in which small areas prone to fires were also notified. The research was moved to a grid-based structure (5 km $\times$ 5 km) in 2017, enabling parameters to be quantified and interpreted within these grids. Detailed additional parameters were also included to make the "Early Warning Alert System for Forest Fire" more effective.

3.4 Forest Fire Alert System Ver. 3.0 (FAST 3.0)

In January 2019, the FSI Forest Fire Alarm underwent a refurbishment and was a quicker, faster, and more stable variant of the earlier fire alert systems. It featured

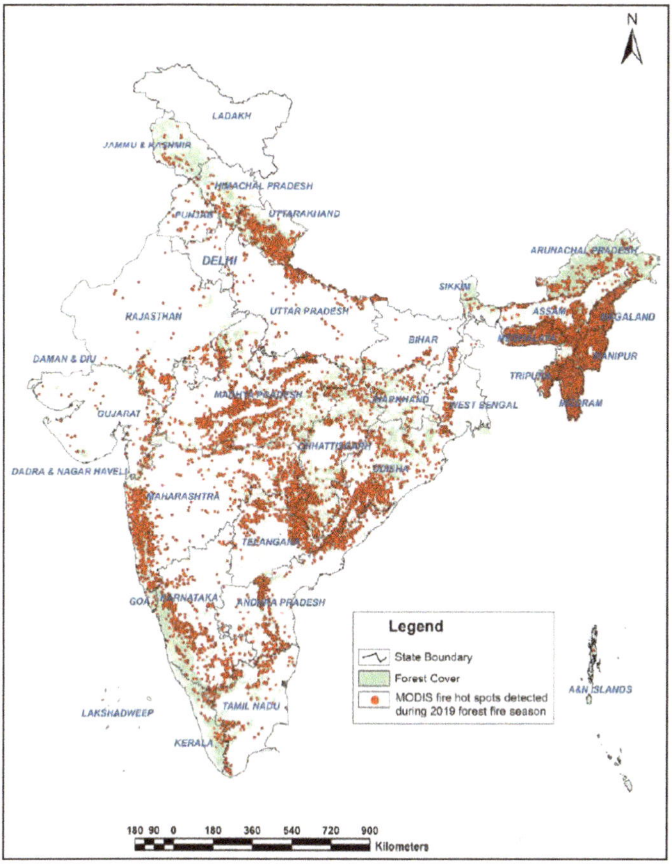

Fig. 1 MODIS-based fire hotspots detected during 2019 forest season in the whole of India

an extensive Forest Fire Monitoring Program: This is based on satellite data (SNPP-VIIRS) to detect and track major forest fire events automatically; FSI Forest Fire Geoportal: together with other thematic layers, to display forest fire-related data; and Web Map Service (WMS): open to the State Forest Departments for incorporation. Customized alerts for 20 beat-level states and two range-level states. Enhanced Method of feedback (via SMS and nodal officer page).

3.5 The Himalayan Region of Jammu and Kashmir—Forests and Forest Fires

The UT of Jammu and Kashmir, India's northernmost region, is renowned for its biodiversity. It is blessed with rich fauna and flora that thrive in its diverse natural habitats. It covers beautiful forests and a 42,549.87 km area and is located between 73° 44′ 58.126"E-76° 46′ 13.099"E and 32° 16′ 48.475"N- 34° 49′ 30.86"N.

Forests are vital to human existence on earth. Jammu and Kashmir state is richly endowed with various forest resources that also serve as catchments for significant Himalayan rivers. Forests are essential for soil conservation, water security, timber, firewood, fodder, and other forest products to meet local people's needs. Being the union territory's largest land-based resource, forests have immense potential to support livelihoods and alleviate poverty.

Forest fires, including soil inhabitants, are regarded as one of the most significant environmental threats that cause considerable harm to forest fauna and flora (Nathawat et al. 2010). The availability of soil nutrients, the development of plant communities, biological diversity, and the atmosphere are strongly affected by forest fires, resulting in enormous economic effects in the region (Rather et al. 2018). The incidence of forest fires can also affect human health adversely. Vast quantities of pollutants, such as particulate matter, carbon monoxide, nitrogen oxides, sulfur dioxide, and organic compounds, are released by forest fires. Particulate matter has potentially harmful health effects as it penetrates deep into the human lungs, causing respiratory and cardiovascular diseases.

According to official data from Jammu and Kashmir, forest fires have caused an impact of 341 ha in 2016. Various incidents of fire across the country in 2017 affected 2.579 ha of forest area. The total area affected was 341.4 ha, and two hundred and fourteen such incidents occurred across the state in 2015. Fire-related events occurred in Jammu and Kashmir in 2014 and 2013 are 470 and 278. In 2016, 214 forest fire events were happening, while the number in 2017 rose to 775. Officials from the Forestry Department said the rising temperatures and the lack of rainfall were crucial factors behind the forest fire rise. People cause more than 90% of incidents involving forest fires.

In the Jammu region's subtropical forests, forest fires are common, while in the Kashmir region, they are low in intensity and frequency. Factors like low and erratic winter rainfall, dry autumn, and spring create a climate conducive to global climate change. Forest fires are also caused. Forest fires are causing significant environmental, economic, and social problems in various countries, with potential long-term effects on the natural environment and the economy.

The J&K Forest Policy also focuses on forest fires' management aspects and enlisting it as an essential aspect of forest protection. Proper equipment and a trained workforce are envisaged to provide effective management of forest fires in vulnerable areas. Often stressed is the role of local governments in the prevention and control of forest fires.

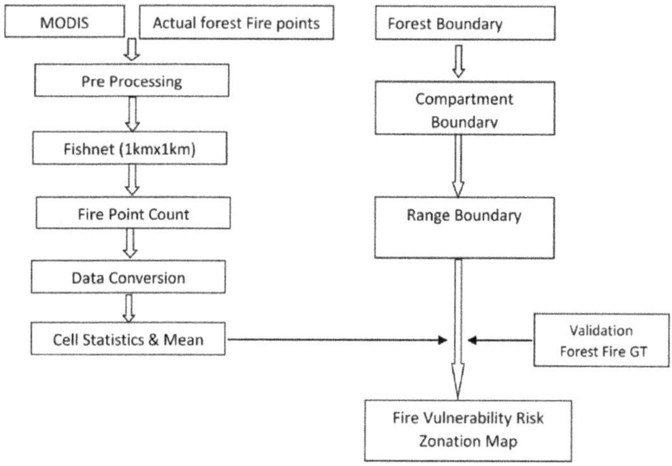

Fig. 2 Methodology flowchart for assessing the forest fire vulnerability risk zonation in the UT of Jammu and Kashmir

3.6 Forest Fire Vulnerability Assessment of UT of Jammu and Kashmir—A Case Study

We assessed the forest fire vulnerability map by combining forest fire points, forest compartment boundary, and range boundaries by utilizing remote sensing and GIS technologies. Hotspot analysis of historical fire data was also performed.

4 Methodology

The overall methodology of this case study is shown in Fig. 2. We obtained point data about actual forest fire incidences from Forest Department for 2002–2018. MODIS satellite fire data (2012–2018) was downloaded from the FIRMS Web site. Fire Information for Resource Management System is a NASA-funded application. Forest Management Boundary comprising divisions, ranges, and compartments was obtained from Forest Department.

5 Results and Discussion

All the actual ground-based forest fire points recorded by the Forest Department during the period (2002–2018) were analyzed in GIS (Fig. 3). The recorded forest fire points (FFPs) numbering 4392 include repeat detections of continuing forest fires also. MODIS hotspot data for the 2012–2018 fires was compared against actual forest

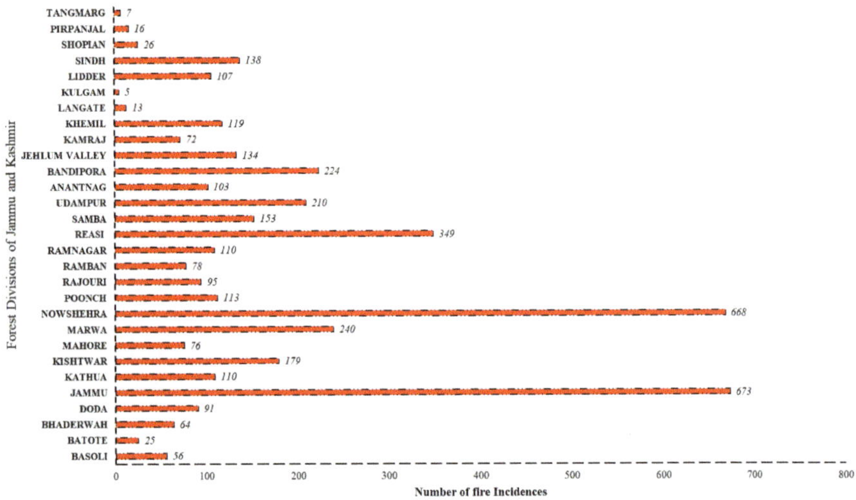

Fig. 3 Division-wise forest incidences in the UT of Jammu and Kashmir (2002–2018)

fire incidence for the corresponding period to identify positives and false negatives. All the detected forest fire points numbering 3141 from MODIS data from 2012 to 2018 were analyzed in GIS. A correlation index of MODIS (3141) and actual fire incidences (2438) from 2012 to 2018 was derived, which showed a correlation coefficient of 0.97 using the seven-year dataset of actual fire points (Fig. 4). Since

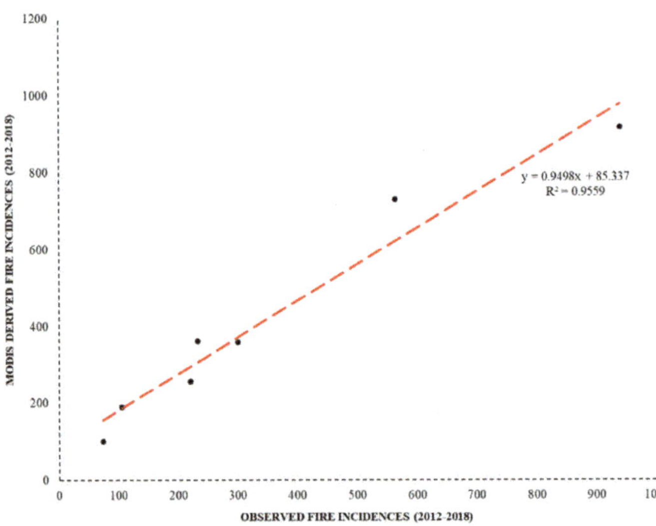

Fig. 4 Regression analysis between actual and MODIS-based fire incidences between 2012 and 2018 in the UT of Jammu and Kashmir

the points derived from MODIS satellite data cover a larger extent, the number of points varies with actual forest fire points.

Both incidences-based and MODIS-derived points were merged, and a total of 8740 points were available for analysis. The shapefile containing vector points of hotspots was projected in ArcGIS. All shapefiles were added thereafter to ArcGIS and projected using the UTM WGS-84 coordinate reference system to ensure spatial accuracy.

Fishnets were created of 1 km × 1 km. The analysis was done on the premise that a fire-prone forest area will show a relatively higher number of detected forest fire points over a long time, i.e., 17 years. The frequency of forest fire points in each grid of 1 km × 1 km was determined through the GIS analysis. In the next step, points were calculated in each grid using cell statistics, which computes per cell (grid) statistics from multiple layers. The cell statistics tool is an experimental method to identify clusters and assumes event density at any location. Available statistics are mean, majority, range, deviation, maximum, sum, median, minimum, and minority. The default static type is mean. Our objective was to find the mean of all cells. Each vulnerability map grid was assigned into classification statistics as total count, minimum, maximum, sum, and mean. Based on the derived frequency of FFPs per year, each grid was categorized in terms of fire proneness using the following criteria:

Each division clipped final vulnerability maps and range boundary of the union territory of J&K (Meraj et al. 2016). Each range map of the union territory of J&K has been prepared using the outcome of cell statistics, and each attribute table is exported to Microsoft Excel. Weightage was then assigned to each map based on the number of fire incidences into high, medium, and low (Table 1, Fig. 5).

Changes in the fire dynamics and increase in the frequency of extreme temperature events have led to a situation in which fires are now a significant threat to forests and their biodiversity. In the Jammu region's subtropical forests, forest fires are normal, while their intensity and frequency in the Kashmir region have not been great. However, for the last few years, with global climate changes taking place, precipitation in winter has not been typical, leading to dry autumn, which creates an environment conducive to forest fires. Besides, directly destroying the forest trees, forest regeneration, microclimate, soil erosion, and wildlife are also adversely affected by the fire. In most situations, forest fire causes the forest vegetation to regress.

The total number of forest fire incidents is 4392, and the number of polygons in compartments is 6795. There are 28 total forest divisions, and 102 total forest ranges in J&K. MODIS fire points are 3141(2012–2018). The annual fire pattern analysis found that 2016, 2018, and 2009 are the peak years of forest fire incidences. With the advent of monsoon rains, the number of fire incidences decreases significantly after June, becoming marginal in July and August.

Table 1 Summary of fire incidences in vulnerable zones based on 1 km × 1 km grid

Division Jammu			
Division	High vulnerability (7.4–11)	Medium vulnerability (3.6–7.4)	Low vulnerability (1–3.6)
Basoli		4	59
Batote			33
Bhaderwah			64
Billawar		8	138
Doda		21	66
Jammu	1	148	513
Kathua		16	108
Kishtwar		4	181
Marwah		6	244
Mahore			79
Nowshehra	1	297	427
Poonch		17	108
Ramnagar		18	118
Ramban			79
Reasi		45	272
Samba	1	17	132
Udhampur		38	198
Total	3	639	2819
Division Kashmir			
Anantnag		4	103
Bandipora	3	56	122
Jhelum Valley		8	126
Kamraj		7	65
Khemil	1	15	92
Kulgam			5
Langate			14
Lidder		13	101
Pirpanjal		4	12
Shopian		12	14
Sindh		21	125
Tangmarg		2	6
Total	4	142	785
Grand total	7	781	3604
Total number of incidences UT of Jammu and Kashmir = 4392			

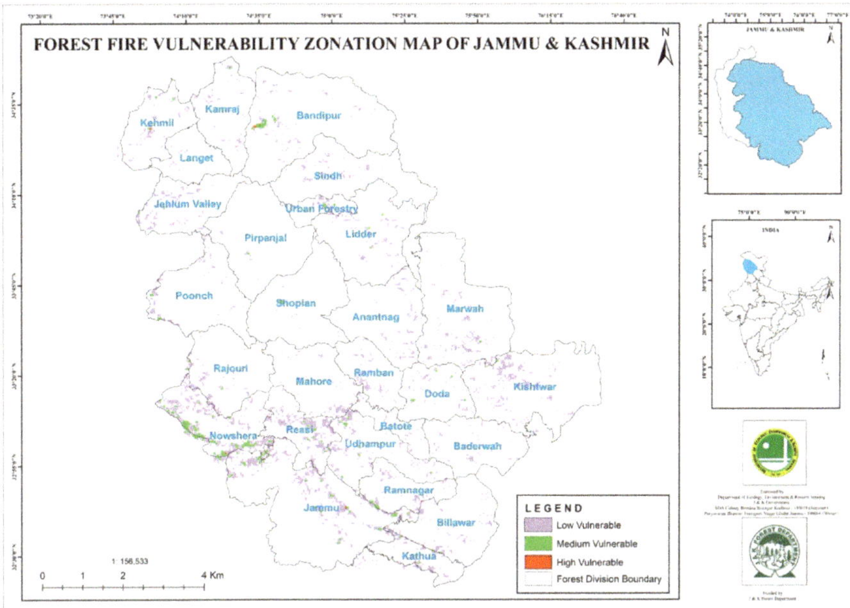

Fig. 5 Forest fire vulnerability map of UT of Jammu and Kashmir

6 Conclusions

The current study uses real forest fire incidences from monsoon rains' arrival (2002–2018). Three divisions are highly vulnerable in the Jammu region, namely Jammu, Nowshehra, and Reasi, with the highest number of fire incidences. Bandipora shows the highest number of forest fire incidences in the Kashmir region division. Forest fire incidents have become frequent in Kashmir under the current weather and climate conditions. In this season, which was an unusual sight in the past, the scanty precipitation in the winter coupled with a spurt rise in day temperature increased forest fires' chances. Forest fires would typically occur in the Kashmir region in the month of May–July, when atmospheric temperature and dryness offer favorable conditions for a fire to begin. The fuel's moisture content influences the rate of combustion, the rate of spread, the probability of ignition, and the reduction of radiant emissions from the flames generated. Even when the fuel moisture content is as high as 25–30%, the rate of fire spread in pine forests increases dramatically, compared to 6–8% in other forests. Factors such as low precipitation, topography, relative humidity, droughts, wind speed or direction, rainfall and sunshine canopy interception, dew deposition, fog, and underlying soil further aggravate this. Due to the sudden temperature rise or when some grazers deliberately throw splinters on the dry grass, resulting in fire, the pine forest is susceptible to fire this season.

Acknowledgements The authors express their gratitude to the Jammu and Kashmir State Forest Department for imbibing us with the task of conducting forest vulnerability assessment of the UT of Jammu and Kashmir.

References

Altaf F, Gowhar M, Romshoo SA (2013) Morphometric analysis to infer hydrological behaviour of Lidder watershed, Western Himalaya, India. Geogr J

Altaf S, Gowhar M, Romshoo SA (2014) Morphometry and land cover based multi-criteria analysis for assessing the soil erosion susceptibility of the western Himalayan watershed. Environ Monitor Assess 186(12):8391–8412

Bhatt CM et al (2017) Satellite-based assessment of the catastrophic Jhelum floods of September 2014, Jammu & Kashmir, India. Geomat Nat Haz Risk 8(2):309–327

Dhyani SK, Handa AK (2013) Area under agroforestry in India: an assessment for present status and future perspective. Indian J Agrofor 15(1):1–11

FSI (Forest Survey of India) (2015) India state of forest report

Gujree I et al (2017) Evaluating the variability and trends in extreme climate events in the Kashmir Valley using PRECIS RCM simulations. Modeling Earth Syst Environ 3(4):1647–1662

Joy J, Shruti K, Singh SK (2019) Kerala flood 2018: flood mapping by participatory GIS approach, Meloor Panchayat. Int J Emerg Techn 10(1):197–205

Kanga S, Singh SK, Sudhanshu (2017) Delineation of urban built-up and change detection analysis using multi-temporal satellite images. Int J Recent Res Aspects 4(3):1–9

Kanga S, Singh SK (2017b) Forest fire simulation modeling using remote sensing & GIS. Int J Adv Res Comput Sci 8(5)

Kanga S et al (2020) Modeling the spatial pattern of sediment flow in lower Hugli Estuary, West Bengal, India by quantifying suspended sediment concentration (SSC) and depth conditions using geoinformatics. Appl Comput Geosci 100043

Kanga S, Kumar S, Singh SK (2017a) Climate induced variation in forest fire using remote sensing and GIS in Bilaspur District of Himachal Pradesh. Int J Eng Comput Sci 6(6):21695–21702

Kumar S, Singh SK, Mahendra RS (2018) Assessment of land use and land cover during pre and post cyclone Phailin in Southern Part of Odisha Coast. Manager's J Future Eng Technol 13(4):53

Meraj G et al (2018b) An integrated geoinformatics and hydrological modelling-based approach for effective flood management in the Jhelum Basin, NW Himalaya. In: Multidisciplinary digital publishing institute proceedings, vol 7(1), 8

Meraj G et al (2015) Assessing the influence of watershed characteristics on the flood vulnerability of Jhelum basin in Kashmir Himalaya. Nat Hazards 77(1):153–175

Meraj G et al (2018a) Geoinformatics based approach for estimating the sediment yield of the mountainous watersheds in Kashmir Himalaya, India. Geocarto Int 33(10):1114–1138

Meraj G, Romshoo SA, Altaf S (2016) Inferring land surface processes from watershed characterization. Geostatistical and geospatial approaches for the characterization of natural resources in the environment. Springer, Cham, 741–744

MoEFCC (2018) Integrated e-governance portal for automation, streamlining, and effective management of processes related to plantation and other forestry work. http://egreenwatch.nic.in/. Accessed June 20 2018

Nathawat MS et al (2010) Monitoring & analysis of wastelands and its dynamics using multiresolution and temporal satellite data in part of Indian state of Bihar. Int J Geomatics Geosci 1(3):297–307

Rather MA et al (2018) Remote sensing and GIS based forest fire vulnerability assessment in Dachigam National park, North Western Himalaya. Asian J Appl Sci 11(2):98–114

Rather MA et al (2017) Assessing the influence of watershed characteristics on soil erosion susceptibility of Jhelum basin in Kashmir Himalayas. Arab J Geosci 10(3):59

Roy PS (2003) Forest fire and degradation assessment using satellite remote sensing and geographic information system. Satellite remote sensing and GIS applications in agricultural meteorology, 361

Hazard Mitigation and Climate Change in the Himalayas–Policy and Decision Making

Mohammd Rafiq, Gowhar Meraj, Amit Parashuram Kesarkar, Majid Farooq, Suraj Kumar Singh, and Shruti Kanga

Abstract Climate change has drastic impacts on the patterns of precipitation across the world. These changes have greatly affected the mountainous regions where the frequency of disasters have increased rapidly. Flash floods, Glacier Lake outburst floods (GLOF), Snow avalanches, Landslides, and Landslide Lake Outburst floods (LLOF) all are increasing in the number as well as in the magnitude of their adverse consequences. One phenomenon describes the climate of one region and the other has dominance at other parts of the region. However, the precipitation over the Indian region hinge on the monsoon, where the southwest monsoon has a major role as it dominates a larger area. On the other hand Himalayan region is mostly dominated by the western disturbances. Precipitation extremes have increased by almost 77.52% per unit increase in temperature over India from last 113 years which have resulted in the surge of disasters. Air temperatures show increasing trends over the Himalayan region, and are consistent with the decrease in the snowfall. In the Himalayan region, approximately 25% decrease in snowfall has been linked with the increase in the minimum temperatures. This has greatly influenced the snow depth, and as such there is 71 ± 24% decline in it in the region. Warming due to climate change has dire consequences, over the Himalayas. The increased occurrences of extreme/heavy precipitation events due to warming climate is the most impact-relevant consequence as it directly impacts the sustainability of these regions. The climate models predict intensification of extreme rain events including

M. Rafiq · A. P. Kesarkar
Department of Space, Government of India, National Atmospheric Research Laboratory, Gadanki 517112, India

G. Meraj (✉) · M. Farooq · S. Kanga
Centre for Climate Change and Water Research (C3WR), Suresh Gyan Vihar University, Jaipur, Rajasthan 302017, India

G. Meraj · M. Farooq
Department of Ecology, Environment and Remote Sensing, Government of Jammu and Kashmir, Kashmir 190018, India

S. K. Singh
Centre for Sustainable Development, Suresh Gyan Vihar University, Jaipur, Rajasthan 302017, India

S. Kanga et al. (eds.), *Disaster Management in the Complex Himalayan Terrains*, Geography of the Physical Environment,
https://doi.org/10.1007/978-3-030-89308-8_12

intensification of cyclones. So, estimating the precipitation precisely is of utmost importance.

Keywords Precipitation extremes · Cryosphere · Climate change · Mountain hazards · Remote sensing · Himalayas

1 Introduction

The heavy precipitation has evolved many disasters including Flash floods, Glacier Lake outburst floods (GLOF), Snow avalanches, Landslides, Landslide Lake Outburst floods (LLOF), etc., (Bhatt et al. 2017). As the climate of India is dominated by different phenomenon at different places and the climate varies at space and time (Joy et al. 2019; Kanga et al. 2020). One phenomenon describes the climate of one region and the other has dominance at other parts of the region. But mostly the precipitation of India depends on the monsoon, where southwest monsoon has a major role, as it dominates a larger area. Also, the westerlies or western disturbances have their role in the top most northern part of India where precipitation is in the form of snow. The heavy precipitation is increasing from last 113 years. The increase is about 62.71% ± 20.32%, 77.52% ± 28.41% and 59.76% ± 20.32% per unit increase in temperature over Western, North Eastern and Central India, respectively. From last 113 years (1901–2013), the increase in the temperature is 0.08 K per decade. It is also reported that in the Central India heavy precipitation is increasing by 4.8%. An increase of about 52% in the heavy precipitation over the last 113 years per degree rise in temperature is projected (Fig. 1). The frequency of heavy rainfall events over India from 1901to 2013 has increased by almost 3.72% per decade. These changes in the precipitation patterns have trigged many disasters. They have large scale effects on the Himalayan regions as these are responsible for GLOF (Ives et al. 2010; Rafiq et al. 2019; Kaushik 2020), Landslides (Dortch et al. 2009), Flash floods (Rasmussen and Houze 2012), LLOF (Ruiz-Villanueva et al. 2017), Snow avalanches (Parshad et al. 2019), etc. In this book chapter, a number of such cases are explored, and the impacts of the changing precipitation patterns over Himalayas are explored. We have also examined the changes in precipitation on vital resources like snow and glaciers in the below sections some of the case studies will be briefed upon.

2 Recent Changes in the Snowfall Patterns Due to Climate Change and Its Impacts on the Himalayas

In the month of January 2017, the winter season of 2019, 2020 and 2021 multiple heavy snowfall events occurred over Jammu and Kashmir which triggering many avalanches and destroyed life and property. The annual snowfall shows a decreasing trend over Jammu and Kashmir from last 40 years but there are increases in the

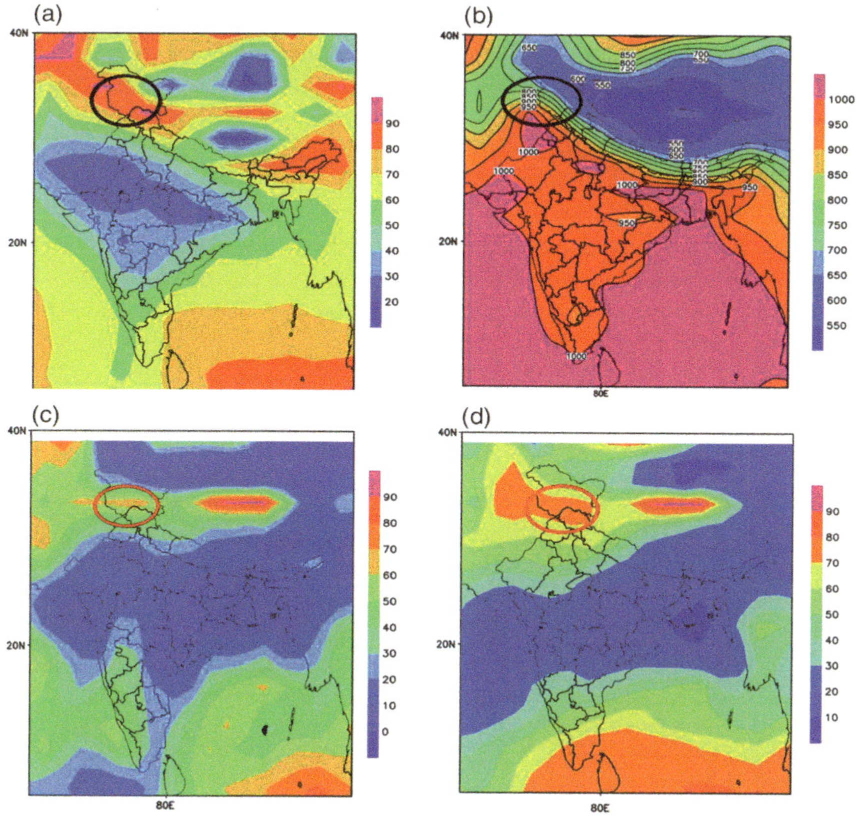

Fig. 1 Shows the **a** relative humidity, **b** pressure (hPa), **c** cloud, **d** cloud cover (%) during January 2017. *Source* Rafiq and Mishra (2018)

extreme snowfall event from last decade. These extremes are linked to the climate change over the north-western Himalayas. Many researchers have quantified the impact of temperature on precipitation where they have noted the increase the extreme precipitation is linked with the increase in the temperature. This temperature increase leads to changes in precipitation patterns (Trenberth et al. 2003, O'Gorman 2012; Villarini et al. 2013; Altaf et al. 2013, 2014; Mishra and Liu 2014, Donat et al. 2016). Increased temperatures lead to an increase in the water keeping potential of the atmosphere, and an increase in the amount of water vapour in the atmosphere could, in fact, result in an increase in precipitation both rainfall as well as Snowfall intensity due to the release of extra latent heat. Constant intense snowfall over a complex topographical region force avalanches, causing severe economic losses. In 1895, Jammu and Kashmir experienced a catastrophic avalanche, resulting in a major loss of life and land which destroyed 500 houses and killed more than 150 people (Lawrence 1895; Meraj et al. 2015a, b; Meraj et al. 2016).

Jammu and Kashmir receives the snowfall during the winter months and is often caused by the Western Disturbances (WD). These WD which originate over Mediterranean Sea and travel eastwards to the Valley of Kashmir where they start precipitating as the convective system in these WD is developed in the upper atmosphere it travels a large distance. Finally when it comes in contact with the Himalayas it starts precipitating. Sometimes the moisture is supplied to this system from Arabian Sea also (Rafiq et al. 2021).

Relative humidity, pressure anomaly and the cloud cover data from National Centres for Environmental Prediction (NCEP)/National Centre for Atmospheric Research (NCAR) was analysed for the month of January 2017 (Rafiq and Mishra 2018) to study the heavy snowfall event of 2017 over the Union territory of Jammu and Kashmir. They reported the formation of dense clouds over the north-western Himalaya due to a low pressure and increase in the water vapour (Fig. 1). Intense storms formed due to WD over the north-western Himalayas, resulting in extreme snowfall over the Jammu and Kashmir region.

Multiple snowfall events occurred during 6th to 8th, 15th to 17th and 24th to 27th of January 2017 over the Jammu and Kashmir region. Heavy snowfall events were recorded over Pahalgam, Gulmarg, Kukernag, Qazigund, Banihall and Kupwara (Fig. 2).

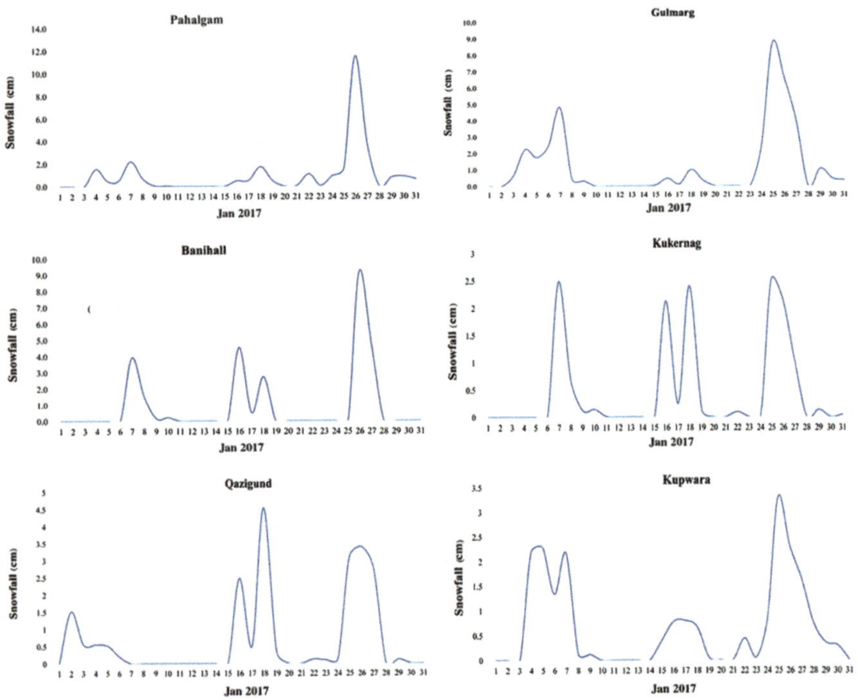

Fig. 2 Snowfall over Pahalgam, Gulmarg, Kukernag, Qazigund, Banihall and Kupwara. *Source* Rafiq and Mishra (2018)

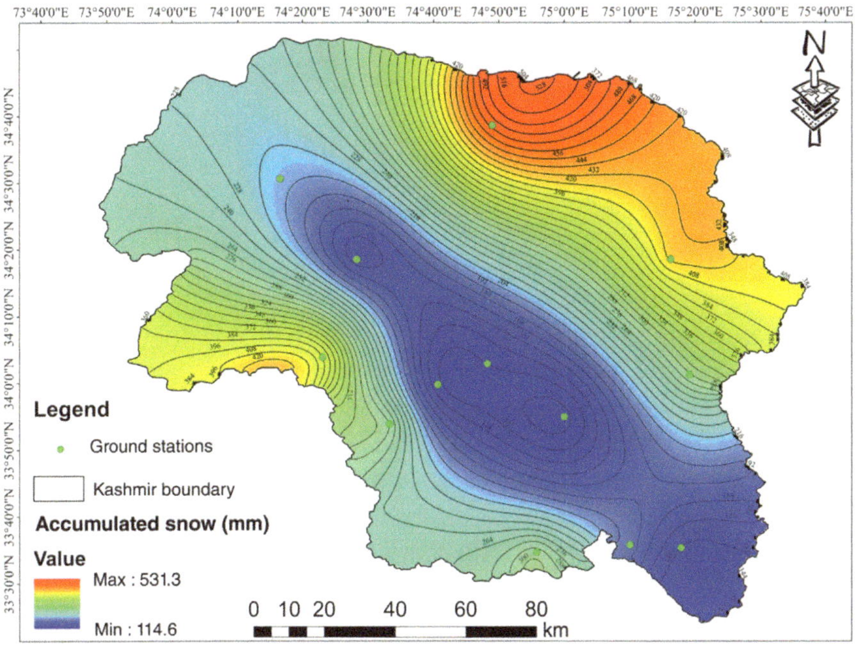

Fig. 3 Snowfall accumulated over Kashmir valley during the month of January 2017. *Source* Rafiq and Mishra (2018)

The accumulated snowfall during the month of January 2017 is shown in Fig. 3. The maximum snowfall has occurred over the Drass area and also the snow avalanches were reported from this area. With the complex topography Kashmir Valley has the elevation range between 1800 and 5300 m AMSL. The lapse rate plays an important role on the snowfall distribution. With the large variations in the lapse rate ~12 C/Km (Rafiq et al. 2014, 2016; Romshoo et al. 2018) heavy snowfall occurs over the mountains regions having the steep slopes. With higher lapse rates, mountainous areas are colder, and so more precipitation occurs as snowfall and less precipitation accumulates as rainfall. Most of the area of Kashmir valley falls under the zone 5 of the seismic zones of India which means there are more chances of snow avalanches. Quite a heavy snowfall has been observed over the Union territory of Jammu and Kashmir region almost 85% of the land in this region has a slope between 15 and 55°, and these areas are also particularly prone to avalanches.

Huge amounts of snowfall over mountain ranges with steep slopes culminated in a catastrophic snow-avalanche that swept the Indian army camp located in the Drass area and killed more than 15 army personnel stationed in that area. Figure 4 shows the graphical abstraction of this snow-avalanche.

In the winter season of 2019 to 2020, heavy snowfall events were recorded over the Jammu and Kashmir starting from 2019 November. The snowfall was so extreme; it broke all the previous records for single day snow accumulation at some of the

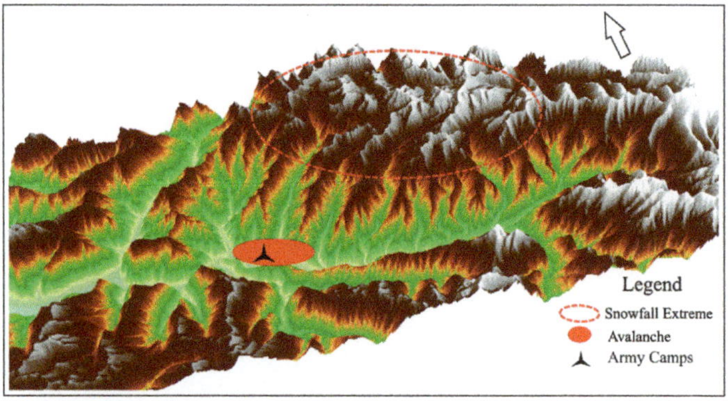

Fig. 4 Showing the location of army camp (star) the avalanche run out zone (Red ellipse) and the heavy snowfall zone (Dotted ellipse)

places. This year's snowfall destroyed the horticulture plantation spread across the Kashmir valley. On the other hand the winter season of 2021 broke all the records of minimum temperatures across the Kashmir valley, the world famous Dal Lake froze and people were seen playing on its surface.

3 Climate Change, Atmosphere and the Cryosphere Interactions

Snowfall across the Jammu and Kashmir has significantly decreased, however, over the past few years extreme snowfall events are happing incessantly. Mishra and Rafiq (2017a, b) have quantified the changes in the snowfall over the Kashmir valley. The snowfall and other data from 1980 to 2010 was analysed for the quantification with temperature and other parameters. It was found that there is a decrease in the snowfall with increasing minimum and maximum temperature over the valley of Kashmir. There is a shift in the snowfall over the years and during last 31 years it was found that the snowfall mostly occurs in the month of February (Fig. 5).

There are numerous snowfall highs and lows across the Kashmir valley. These snowfall increases are consistent with the associated temperature decreases (minimum and maximum). This indicates that the temperature increase allows the snowfall to decline. It is recorded that snowfall over Pahalgam indicates a diminishing pattern in the month of November, December, January, and March. A substantial decadal decrease of about 48 mm in snowfall over Pahalgam in the month of March is observed. This decline in snowfall over Pahalgam is associated with a decadal rise of around 0.7 and 1.90 °C in minimum and maximum temperature (Fig. 6). There is almost 45 mm decline of snowfall over Pahalgam from the last 30 years and about 6 mm over the Gulmarg region which is in consistent with the increases in the temperature over the Pahalgam and the Gulmarg. The large scale changes over the

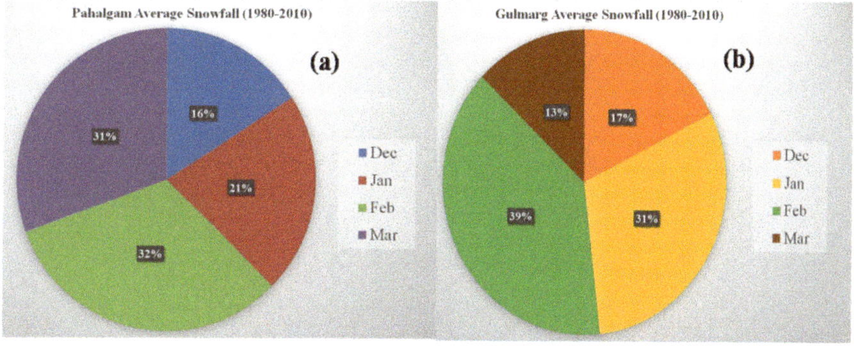

Fig. 5 Displaying the snowfall distribution pie chart from 1980 to 2010 over Pahalgam (**a**), and Gulmarg (**b**)

Pahalgam region may be associated with the changes in the WD over the region as the Pahalgam falls in the Himalaya region and is mostly effected by the WD while as the Gulmarg receives most of the precipitation due to monsoons.

Black carbon concentration (CBCC) over Gulmarg and Pahalgam shows large inter-annual variability. A significant rising trend of 1.3 and 2.8 mg/m^2 per decade is observed over Pahalgam and Gulmarg, respectively, (Fig. 7). Bhat et al. (2017) also reported the increase in the BC concentration over the Kashmir valley.

Inter-annual technique was used to quantify the impacts of warming on the snowfall over the union territory of Jammu and Kashmir. It was found that a unit degree increase in the temperature (minimum) decreases the snowfall by 24% (with an error of 9%) over the Great Himalaya range of mountains. Also, the Pi-Panjal range shows a decline of almost 10% with a unit degree increase in the temperature (Fig. 8).

Glaciers are very essential for a large population of India. They provide additional water supply apart from the precipitation due to monsoons (Pall et al. 2019; Romshoo et al. 2020). They can indicate climate change as they respond to changing temperature and precipitation (Gujree et al. 2017; Kanga et al. 2017a, b). In the last many years studying glaciers of the Himalayas and reading the scientific literature, it can be fairly said that the snowfall variability is responsible for negative mass balance of glaciers which in turn is linked to the changes in the atmosphere. Air temperatures show increasing trends over the Himalayan region, which are consistent with the decrease in the snowfall. If we look at the numbers, approximately a 25% decrease in snowfall is linked with a per degree increase in the minimum temperature. This can influence the snow depth there is 71 $\pm$ 24% decline in the snow depth for each degree increase in temperature observed over the Kolahaoi glacier (Fig. 9). We notice the influence of these trends on glaciers in terms of their shrinkages and negative mass balances. For instance, Jammu and Kashmir‘s largest glacier (Kolahoi Glacier) has shrunk 75% (Fig. 10) of its area since 1857 (Rafiq and Mishra 2017a, b). Glaciers in the Himalayan-Karakoram ranges show a negative mass balance of approximately half a metre per year (water equivalent) since 1975 (Li et al. 2008; Romshoo et al. 2015; Azam et al. 2018). Few studies indicated that the glacier area shrinkage is linked

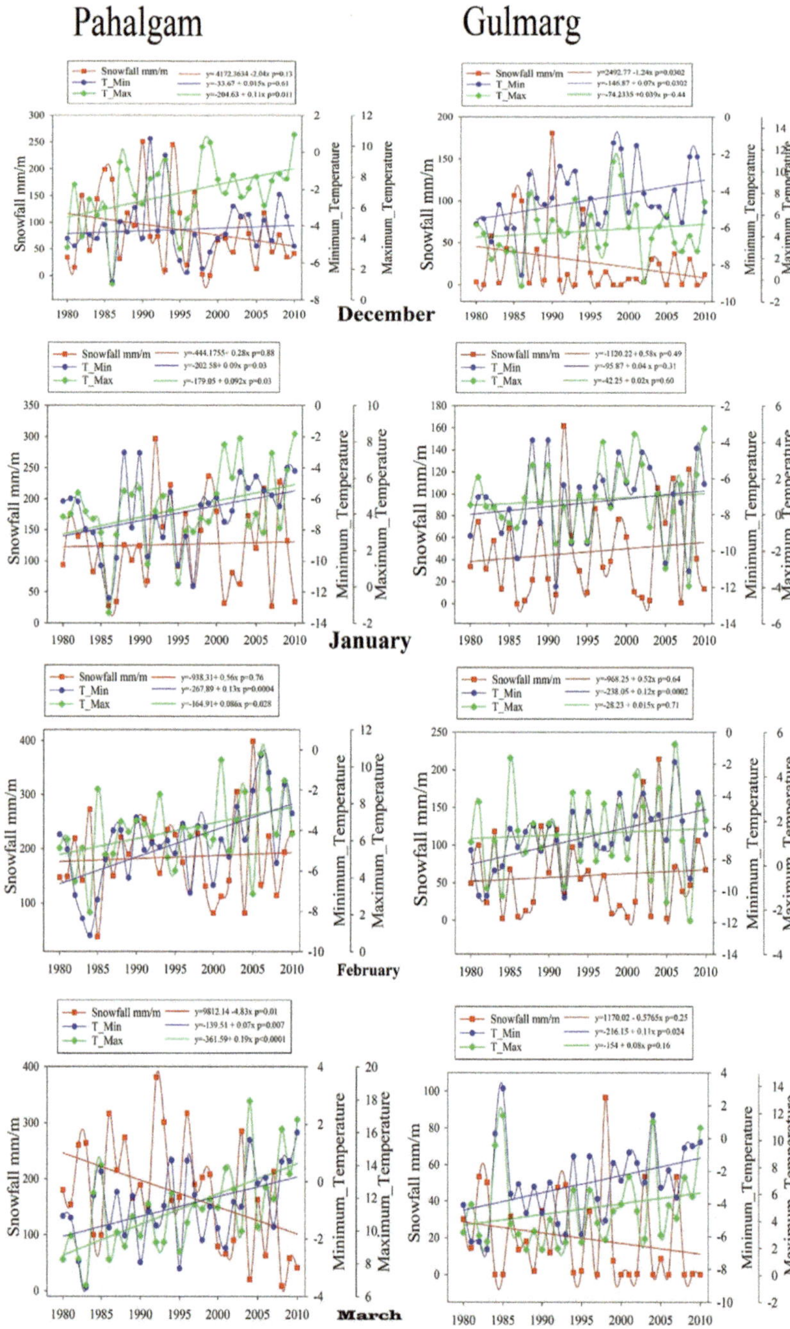

Fig. 6 Displaying the changes in snowfall, minimum and maximum temperature over Pahalgam (**a**), and Gulmarg (**b**)

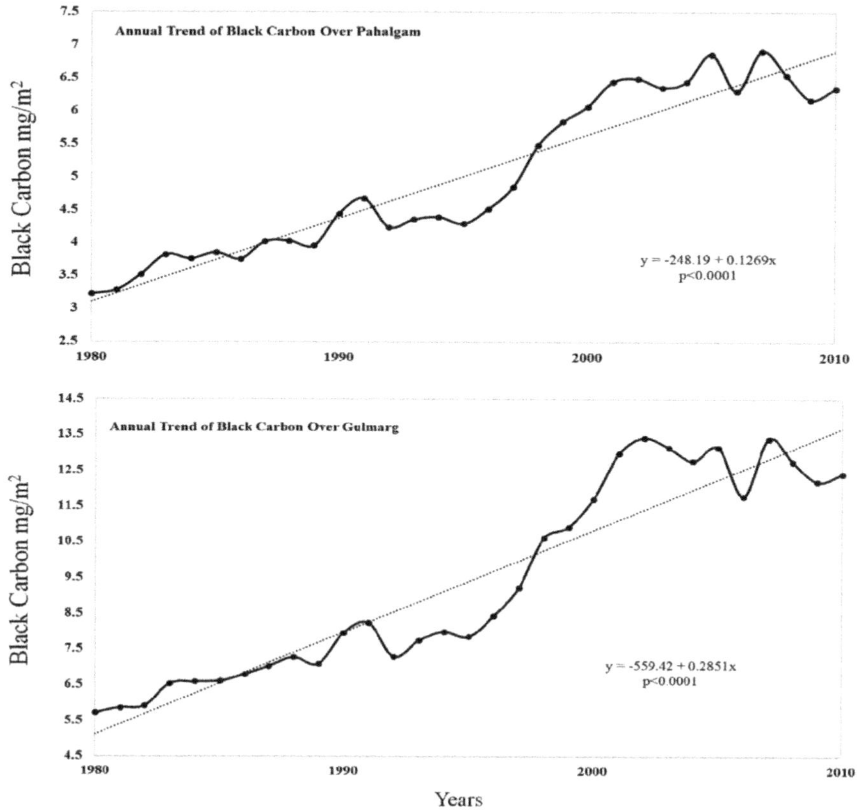

Fig. 7 Black Carbon variations over Pahalgam and Gulmarg. *Source* Mishra and Rafiq (2017a, b)

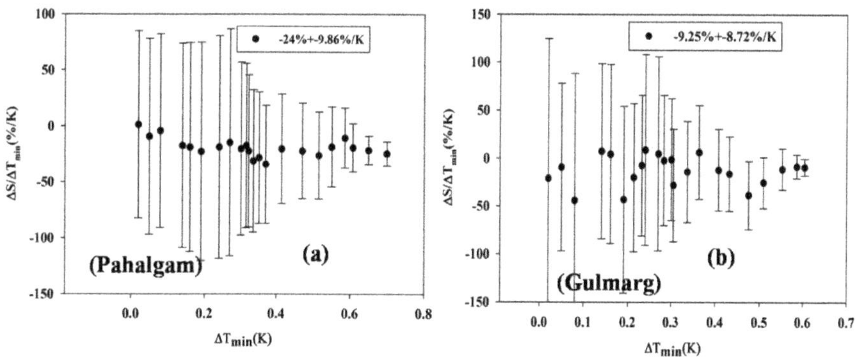

Fig. 8 Per cent changes in the snowfall over Pahalgam (**a**), and Gulmarg (**b**) as a function of temperature

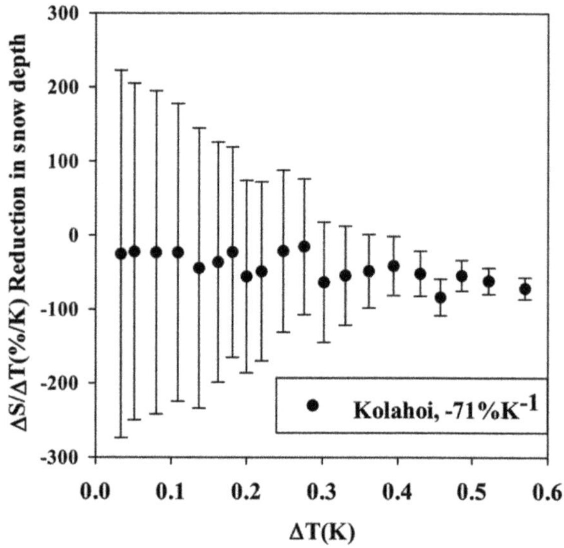

Fig. 9 Snow depth changes over Kolahoi glacier. *Source* Rafiq and Mishra (2016)

to a reduction in snow depth which in turn is affected by the increase in black carbon concentration, temperature, and reduction in the precipitation. The high concentration of Black Carbon observed over the high-altitude Himalayan Kashmir region has serious implications for the regional climate, hydrology and cryosphere which need further investigation (Bhat et al. 2017).

4 Conclusions and the Way Forward for Policy and Decision Making for Adapting to Climate Change

Warming due to climate change has dire consequences, worldwide the changes in extreme/heavy precipitation due to warming is the most impact-relevant consequence. The climate models predict intensification of extreme rain events including intensification of cyclones. So, estimating the precipitation precisely is of utmost importance. The research in monitoring environment from space has made remarkable progress in last 20 years. As we get lot of data, the aim of this researchers is to make a suitable usage of this data. The rapid development in this field provide an opportunity for better and accurate precipitation monitoring and estimating (Mishra and Rafiq 2017a, b). The knowledge of energy cycle and water cycle is increased by the observations in VIS/IR and microwave measurements. All these improvements in space technology provide an opportunity to estimate the precipitation at fine scale and in real-time.

Analysis of precipitation, temperature, and aerosols (like Black Carbon) data have important implications over the Himalayas. Results reveal that there is high

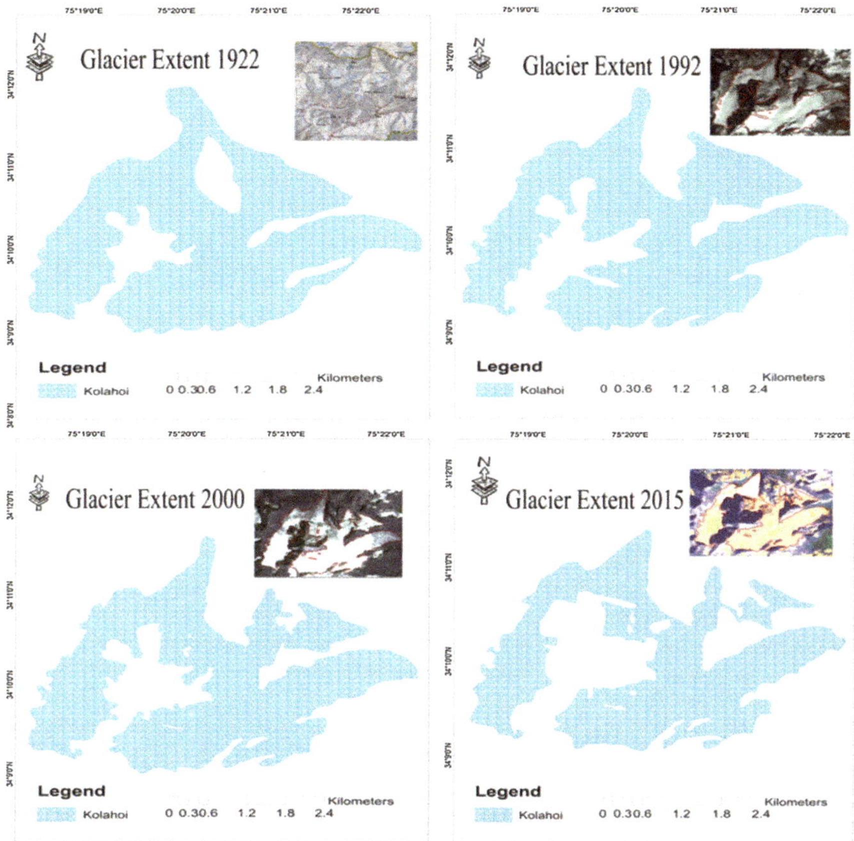

Fig. 10 Dimensional changes in Kolahoi glacier from 1992 to 2015. *Source* Rafiq and Mishra (2016)

inter-annual variability in snowfall for different months over the Himalayas. Great Himalaya region shows a significant reduction in snowfall for the peak snowfall month which is consistent with increase in temperature. This may affect the economy of these regions as tourism is one of the major sources of income (Dar et al. 2014; Rafiq et al. 2018). Increase in temperature is attributed to increase in column Black carbon as a result of unplanned urbanization. Increased temperature may cause significant snowmelt over Himalayan region (Bahuguna et al. 2007). Reduction in snowfall combined with increase in temperature, snowmelt and black carbon concentration may result in shrinking of glaciers (Rafiq and Mishra 2016). Recently Jammu and Kashmir experienced a heavy flood in 2014 which was attributed to heavy precipitation (in the form of rain) and snowmelt from high elevated regions including Pahalgam and Gulmarg (Mishra 2015). Given the inaction in mitigating anthropogenic warming such events may increase in near future. For this reason, water resource management and disaster preparedness need to be implemented at earliest.

The Jammu and Kashmir Government have formulated the State Action Plan on Climate Change (SAPCC J&K). Its main objective is to strategize adaptation and mitigation initiative towards emission stabilization and enhancement of ecosystem resilience, climate proofing of the livelihood sector and diversification of dependency on the natural resources. Considering the issues related to the impacts of climate change on the ecologically sensitive as well as economically important sectors, ten missions specific to the state were identified along with corresponding ten working groups and sanctioned by the Government. These ten missions are, Energy—Solar Mission and Renewable Energy, Enhanced Energy Efficiency, Water, Sustainable Habitat, Sustainable Agriculture, Tourism, Sustainable Himalayan Ecosystem, Health, Disaster Management, and Strategic Knowledge Mission. These missions shall pave way for the better adaptation strategies to cope up with climate change in Kashmir Himalayas.

References

Altaf F, Meraj G, Romshoo SA (2013) Morphometric analysis to infer hydrological behaviour of Lidder watershed, Western Himalaya, India. Geogr J

Altaf S, Gowhar M, Romshoo SA (2014) Morphometry and land cover based multi-criteria analysis for assessing the soil erosion susceptibility of the western Himalayan watershed. Environ Monitor Assessm 186(12):8391–8412

Azam MF, Wagnon P, Berthier E, Vincent C, Fujita K, Kargel JS (2018) Review of the status and mass changes of Himalayan-Karakoram glaciers. J Glaciol 64(243):61–74

Bahuguna IM, Kulkarni AV, Nayak S, Rathore BP, Negi HS, Mathur P (2007) Himalayan glacier retreat using IRS 1C PAN stereo data. Int J Remote Sens 28(2):437–442

Bhat MA, Romshoo SA, Beig G (2017) Aerosol black carbon at an urban site-Srinagar, Northwestern Himalaya, India: seasonality, sources, meteorology and radiative forcing. Atmos Environ 165:336–348

Bhatt CM et al (2017) Satellite-based assessment of the catastrophic Jhelum floods of September 2014, Jammu & Kashmir, India. Geomat Nat Haz Risk 8(2):309–327

Dar RA, Rashid I, Romshoo SA, Marazi A (2014) Sustainability of winter tourism in a changing climate over Kashmir Himalaya. Environ Monit Assess 186(4):2549–2562

Donat MG, Lowry AL, Alexander LV, O'Gorman PA, Maher N (2016) More extreme precipitation in the world's dry and wet regions. Nat Clim Chang 6(5):508–513

Dortch JM, Owen LA, Haneberg WC, Caffee MW, Dietsch C, Kamp U (2009) Nature and timing of large landslides in the himalaya and Trans himalaya of Northern India. Quatern Sci Rev 28(11–12):1037–1054

Gujree I et al (2017) Evaluating the variability and trends in extreme climate events in the Kashmir Valley using PRECIS RCM simulations. Model Earth Syst Environ 3(4):1647–1662

Ives JD, Shrestha RB, Mool PK (2010) Formation of glacial lakes in the Hindu Kush-Himalayas and GLOF risk assessment. ICIMOD; May, Kathmandu

Joy J, Shruti K, Singh SK (2019) Kerala flood 2018: flood mapping by participatory GIS approach, Meloor Panchayat. Int J Emerg Techn 10(1):197–205

Kanga S, Singh SK (2017b) Forest fire simulation modeling using remote sensing & GIS. Int J Adv Res Comput Sci 8(5)

Kanga S et al (2020) Modeling the spatial pattern of sediment flow in lower Hugli Estuary, West Bengal, India by quantifying suspended sediment concentration (SSC) and depth conditions using geoinformatics. Appl Comput Geosci 100043

Kanga S, Kumar S, Singh SK (2017a) Climate induced variation in forest fire using remote sensing and GIS in Bilaspur District of Himachal Pradesh. Int J Eng Comput Sci 6(6):21695–21702
Kaushik S, Rafiq M, Joshi PK, Singh T (2020) Examining the glacial lake dynamics in a warming climate and GLOF modelling in parts of Chandra basin, Himachal Pradesh, India. Sci Total Environ 714:136455
Lawrence WR (1895) The valley of Kashmir. Asian Educational Services, New Delhi
Li X, Cheng G, Jin H, Kang E, Che T, Jin R, Wu L, Nan Z, Wang J, Shen Y (2008) Cryospheric change in China. Global Planet Change 1, 62(3–4):210–218
Meraj G, Romshoo SA, Yousuf AR, Altaf S, Altaf F (2015) Assessing the influence of watershed characteristics on the flood vulnerability of Jhelum basin in Kashmir Himalaya: reply to comment by Shah 2015. Nat Hazards 78(1):1–5
Meraj G et al (2015a) Assessing the influence of watershed characteristics on the flood vulnerability of Jhelum basin in Kashmir Himalaya. Nat Hazards 77.1:153–175
Meraj G, Romshoo SA, Altaf S (2016) Inferring land surface processes from watershed characterization. Geostatistical and geospatial approaches for the characterization of natural resources in the environment. Springer, Cham, 741–744
Mishra A, Liu SC (2014) Changes in precipitation pattern and risk of drought over India in the context of global warming. J Geophys Res Atmos 119(13):7833–7841
Mishra A, Rafiq M (2017) Towards combining GPM and MFG observations to monitor near real time heavy precipitation at fine scale over India and nearby oceanic regions. Dyn Atmos Oceans 80:62–74
Mishra AK (2015) A study on the occurrence of flood events over Jammu and Kashmir during September 2014 using satellite remote sensing. Nat Hazards 78(2):1463–1467
Mishra AK, Rafiq M (2017) Analyzing snowfall variability over two locations in Kashmir, India in the context of warming climate. Dyn Atmos Oceans 1(79):1–9
Meraj G, Romshoo SA, Altaf S (2016) Inferring land surface processes from watershed characterization. In: Geostatistical and geospatial approaches for the characterization of natural resources in the environment. Springer, Cham, pp 741–744
O'Gorman PA (2012) Sensitivity of tropical precipitation extremes to climate change. Nat Geosci 5(10):697–700
Pall IA, Meraj G, Romshoo SA (2019) Applying integrated remote sensing and field-based approach to map glacial landform features of the Machoi Glacier valley. NW Himalaya. SN Appl Sci 1(5):1–11
Parshad R, Kumar P, Srivastva PK (2019) Seismically induced snow avalanches at Nubra-Shyok region of Western Himalaya. India. Nat Hazards 99(2):843–855
Rafiq M, Mishra A (2016) Investigating changes in Himalayan glacier in warming environment: a case study of Kolahoi glacier. Environ Earth Sci 75(23):1–9
Rafiq M, Mishra AK (2018) A study of heavy snowfall in Kashmir, India in January 2017. Weather 73(1):15–17
Rafiq M, Mishra AK, Meer MS (2018) On land-use and land-cover changes over Lidder Valley in changing environment. Ann GIS 24(4):275–285
Rafiq M, Rashid I, Romshoo SA (2014) Estimation and validation of remotely sensed land surface temperature in Kashmir valley. J Himalayan Ecol Sustain Dev 9:1–3
Rafiq M, Rashid I, Romshoo SA (2016) Estimating land surface temperature and its lapse rate over Kashmir Valley Using MODIS Data. In: Geostatistical and geospatial approaches for the characterization of natural resources in the environment. Springer, Cham, pp 723–728
Rafiq M, Romshoo SA, Mishra AK, Jalal F (2019) Modelling Chorabari Lake outburst flood, Kedarnath, India. J Mountain Sci 1;16(1):64–76
Rafiq M, Javid P, Kesarkar AP, Lotus S (2021) A Study of extreme snowfall during 2019 and 2020 across the Kashmir Himalaya. Weather 76. https://doi.org/10.1002/wea.3938
Rasmussen KL, Houze RA (2012) A flash-flooding storm at the steep edge of high terrain: disaster in the Himalayas. Bull Am Meteor Soc 93(11):1713–1724

Romshoo SA, Dar RA, Rashid I, Marazi A, Ali N, Zaz SN (2015) Implications of shrinking cryosphere under changing climate on the stream flows in the Lidder catchment in the Upper Indus Basin, India. Arct Antarct Alp Res 47(4):627–644

Romshoo SA, Rafiq M, Rashid I (2018) Spatio-temporal variation of land surface temperature and temperature lapse rate over mountainous Kashmir Himalaya. J Mt Sci 15(3):563–576

Romshoo SA, Fayaz M, Meraj G, Bahuguna IM (2020) Satellite-observed glacier recession in the Kashmir Himalaya, India, from 1980 to 2018. Environ Monit Assess 192(9):1–17

Ruiz-Villanueva V, Allen S, Arora M, Goel NK, Stoffel M (2017) Recent catastrophic landslide lake outburst floods in the Himalayan mountain range. Prog Phys Geogr 41(1):3–28

Trenberth KE, Dai A, Rasmussen RM, Parsons DB (2003) The changing character of precipitation. Bull Am Meteor Soc 84(9):1205–1218

Villarini G, Smith JA, Vecchi GA (2013) Changing frequency of heavy rainfall over the central United States. J Clim 26(1):351–357